The Curriculum Foundations Project

Voices of the Partner Disciplines

Reports from a series of disciplinary workshops organized by the Curriculum Renewal Across the First Two Years (CRAFTY) subcomitteee of the Committee for the Undergraduate Program in Mathematics (CUPM)

Library of Congress Catalog Card Number 2004100780

ISBN 0-88385-813-4

Printed in the United States of America

Current printing (last digit):
10 9 8 7 6 5 4 3 2 1

The Curriculum Foundations Project

Voices of the Partner Disciplines

Reports from a series of disciplinary workshops organized by the Curriculum Renewal Across the First Two Years (CRAFTY) subcomittee of the Committee for the Undergraduate Program in Mathematics (CUPM)

Edited by

Susan L. Ganter
Clemson University

William Barker
Bowdoin College

Published and Distributed by
The Mathematical Association of America

Preface

Given the impact of mathematics instruction on so many other fields of study—especially instruction during the first two years—there is a need for significant input from the partner disciplines when designing the undergraduate mathematics curriculum. An unprecedented amount of information on the mathematical needs of partner disciplines has been gathered through a series of disciplinary-based workshops known as the *Curriculum Foundations Project.*

Informed by the views expressed in these workshops, the Mathematical Association of America (MAA) has completed an extensive study of the undergraduate program in mathematics. The result is a set of recommendations from the MAA Committee on the Undergraduate Program in Mathematics (CUPM)—*Undergraduate Programs and Courses in the Mathematical Sciences: CUPM Curriculum Guide 2004*—that will assist mathematics departments as they plan their programs through the first decade of the 21st century.

The CUPM subcommittee Curriculum Renewal Across the First Two Years (CRAFTY) organized and ran the Curriculum Foundations Project, gathering input from partner disciplines through a series of eleven workshops held across the country from November 1999 to February 2001. A final summary conference, attended by representatives of all the previous workshops, took place in November 2001.

Each Curriculum Foundations workshop consisted of 20–35 participants, the majority chosen from the discipline under consideration, the remainder chosen from mathematics. The workshops were not intended to be discussions between mathematicians and colleagues in the partner disciplines, although this certainly happened informally. Instead, each workshop was a dialogue among the representatives from the partner discipline, with mathematicians present only to listen and serve as resources when questions arose about the mathematics curriculum.

Each workshop produced a report directed to the mathematics community summarizing the workshop's recommendations and conclusions. The reports were written by representatives of the partner disciplines, insuring accurate reporting of the workshop discussions while also adding credibility to the recommendations. The Curriculum Foundations Steering Committee supplied a common set of questions to guide the discussions at each workshop. This helped achieve uniformity in structure for the reports and made it easier to compare the recommendations from different disciplines.

Workshop participants from the partner disciplines were extremely grateful—and surprised—to be invited by mathematicians to state their views about the mathematics curriculum. That the opinions of the partner disciplines were considered important and would be taken seriously in the development of the *CUPM Guide 2004* only added to their enthusiasm for the project as well as their interest in continuing conversations with the mathematics community.

At the final summary conference, held at the U. S. Military Academy at West Point in November 2001, a common set of recommendations—the *Collective Vision*—were distilled from the reports of the previous workshops.

This volume includes CRAFTY's official report on the Collective Vision recommendations as well as the individual disciplinary reports. The disciplinary reports and Collective Vision were used extensively by CUPM to inform the preparation of its *CUPM Guide 2004*. However, the workshop reports and Collective Vision have value independent of the *CUPM Guide 2004* since they also serve as resources for multi-disciplinary discussions at individual institutions. Promoting and supporting informed interdepartmental discussions about the undergraduate curriculum might ultimately be the most important outcome of the Curriculum Foundations Project.

Contents

A Collective Vision

Voices of the Partner Disciplines

Susan L. Ganter
Clemson University

William Barker
Bowdoin College

Introduction

Mathematics can and should play an important role in the education of undergraduate students. In fact, few educators would dispute that students who can think mathematically and reason through problems are better able to face the challenges of careers in other disciplines—including those in non-scientific areas. Add to these skills the appropriate use of technology, the ability to model complex situations, and an understanding and appreciation of the specific mathematics appropriate to their chosen fields, and students are then equipped with powerful tools for the future.

Unfortunately, many mathematics courses are not successful in achieving these goals. Students do not see the connections between mathematics and their chosen disciplines; instead, they leave mathematics courses with a set of skills that they are unable to apply in non-routine settings and whose importance to their future careers is not appreciated. Indeed, the mathematics many students are taught often is not the most relevant to their chosen fields. For these reasons, faculty members outside mathematics often perceive the mathematics community as uninterested in the needs of non-mathematics majors, especially those in introductory courses.

The mathematics community ignores this situation at its own peril since approximately 95% of the students in first-year mathematics courses go on to major in other disciplines. The challenge, therefore, is to provide mathematical experiences that are true to the spirit of mathematics yet also relevant to students' futures in other fields. The question then is not *whether* they need mathematics, but *what* mathematics is needed and in what context.

The Mathematical Association of America (MAA) is currently studying the undergraduate program in mathematics, informed by the views of a broad segment of the mathematics community and its partner disciplines. The goal is a set of recommendations from the MAA Committee on the Undergraduate Program in Mathematics (CUPM)—the *CUPM Curriculum Guide 2004*—that will assist mathematics departments as they plan their programs through the first decade of the 21st century. The current efforts by CUPM are being informed by an unprecedented amount of information on the mathematical needs of partner disciplines, obtained through a series of disciplinary-based workshops known as the *Curriculum Foundations Project*.

The Curriculum Foundations Project

The CUPM subcommittee *Curriculum Renewal Across the First Two Years* (CRAFTY) has gathered input from partner disciplines through a series of eleven workshops held across the country from November 1999 to February 2001, followed by a final summary conference in November 2001 (see Appendix A). Each Curriculum Foundations workshop consisted of 20–35 participants, the majority chosen from the discipline under consideration, the remainder chosen from mathematics. The workshops were not intended to be discussions between mathematicians and colleagues in the partner disciplines, although this certainly happened informally. Instead, each workshop was a dialogue among the representatives from the partner discipline, with mathematicians present only to listen and serve as a resource when questions about the mathematics curriculum arose.

Each workshop produced a report directed to the mathematics community summarizing the workshop's recommendations and conclusions. Each report was written by representatives of the partner discipline, insuring accurate reporting of the workshop discussions while also adding credibility to the recommendations. The Curriculum Foundations Steering Committee supplied a common set of questions to guide the discussions at each workshop (see Appendix B). This helped achieve uniformity in structure for the reports and made it easier to compare the recommendations from different disciplines.

The host institutions funded most of the workshops.[1] Such financial support—obtained with little advance notice—indicates the high level of support from university administrations for such interdisciplinary discussions about the mathematics curriculum. Workshop participants from the partner disciplines were extremely grateful—and surprised—to be invited by mathematicians to state their views about the mathematics curriculum. That the opinions of the partner disciplines were considered important and would be taken seriously in the development of the *CUPM Curriculum Guide 2004* only added to their enthusiasm for the project as well as their interest in continuing conversations with the mathematics community.

In addition to the workshop reports, the Curriculum Foundations Project has resulted in a number of publications that describe the workshops, their outcomes, and related work (see Publications). These publications include articles in journals of the disciplinary societies as well as the general press. Conversations have also continued via panels and invited colloquia at professional meetings, both in mathematics and the partner disciplines. For example, the Joint Mathematics Meetings (January 2001–2004) and MathFests (August 2001–2003) featured a number of events about the Curriculum Foundations Project. In addition, the partner disciplines have enthusiastically continued the conversations from the workshops at their own disciplinary meetings and have included mathematicians in these discussions.

In January 2001, a series of small focus groups was organized by CUPM to discuss the disciplinary workshop reports. Each invited focus group discussed and analyzed the implications of the report(s) for a specific discipline. Input generated from these discussions was used to supplement the workshop reports in the development of the *CUPM Curriculum Guide 2004*.[2]

In November 2001, invited representatives from each disciplinary workshop gathered at the United States Military Academy in West Point, NY for a final Curriculum Foundations Conference. The discussions resulted in *A Collective Vision*, a set of commonly shared recommendations for the first two years of undergraduate mathematics instruction.

This volume includes CRAFTY's official report on the *Collective Vision* recommendations, as well as the individual disciplinary reports.[3] The disciplinary reports and *Collective Vision* were used by CUPM

---

[1] The only exceptions were the two workshops on technical mathematics, which were hosted by two-year institutions and funded by the National Science Foundation.

[2] Information about these and other resources is available on the MAA website (`www.maa.org`) or through the authors.

[3] Electronic versions of these materials are also available for downloading from `www.maa.org/cupm/crafty`.

to inform the preparation of the *CUPM Curriculum Guide 2004*, a set of recommendations for the mathematics community on the undergraduate curriculum that will focus on desired student outcomes. However, the workshop reports and *Collective Vision* have value independent of the *CUPM Curriculum Guide 2004*, since they can serve as resources for multi-disciplinary discussions at individual institutions. Promoting and supporting informed interdepartmental discussions about the undergraduate curriculum might ultimately be the most important outcome of the Curriculum Foundations Project.

Summary Recommendations: A Collective Vision

Understanding, Skills, and Problem Solving

Emphasize conceptual understanding.

- *Focus on understanding broad concepts and ideas in all mathematics courses during the first two years.*
- *Emphasize development of precise, logical thinking. Require students to reason deductively from a set of assumptions to a valid conclusion.*
- *Present formal proofs only when they enhance understanding. Use informal arguments and well-chosen examples to illustrate mathematical structure.*

There is a common belief among mathematicians that the users of mathematics (engineers, economists, etc.) care primarily about computational and manipulative skills, forcing mathematicians to cram courses full of algorithms and calculations to keep "them" happy. Perhaps the most encouraging discovery from the Curriculum Foundations Project is that this stereotype is largely false. Though there are certainly individuals from the partner disciplines who hold the more strict algorithmic view of mathematics, the disciplinary representatives at the Curriculum Foundations workshops were unanimous in their emphasis on the overriding need to develop in students a conceptual understanding of the basic mathematical tools.

The partner disciplines also value the precise, logical thinking that is an integral part of mathematics. They would like to see this emphasized in early collegiate mathematics instruction in a way that enhances understanding of the underlying concepts. However, the partner disciplines vary widely in the level of rigor and formal reasoning their students need to master. For instance, business students do not need much formal mathematical proof, while students in computer science or software engineering must develop the skill to apply formal logic and construct simple but rigorous proofs. It is therefore important that each institution carefully consider the nature of its students and their intended major disciplines when planning how to incorporate mathematical reasoning into the curriculum. Precise thinking and formal proof must be present, but in appropriate courses, at the appropriate time, and delivered in the appropriate manner. When carelessly introduced, formal proof can confuse many students and negatively affect their ability to understand and apply mathematics.

Emphasize problem solving skills.

- *Develop the fundamental computational skills the partner disciplines require, but emphasize* ***integrative skills****: the ability to apply a variety of approaches to single problems, to apply familiar techniques in novel settings, and to devise multi-stage approaches in complex situations.*

Fundamental computational skills are important and must be developed in students. However, colleagues in the partner disciplines confirm that applying mathematics to unfamiliar problems requires far more than computational skill. Mathematics courses must include more sophisticated problem-solving experiences than simple situations in which students look in the book for an example of the same type and change the numbers.

Students must learn to recognize how and when to use mathematics outside of familiar contexts. In the partner disciplines they will have to make judgments about what mathematical techniques (and what technologies) are appropriate for specific problems. Therefore, students must learn more than how to copy the behavior of their instructor; they must develop their own processes for solving problems. Support must be given as students make this transition.

Emphasize mathematical modeling.

- *Expect students to create, solve, and interpret mathematical models.*
- *Provide opportunities for students to describe their results in several ways: analytically, graphically, numerically, and verbally.*
- *Use models from the partner disciplines: students need to see mathematics in context.*

The importance ascribed to mathematical modeling by *every* disciplinary group in *every* workshop was quite striking. After the first two years, students should be able to create, solve, and interpret basic mathematical models from informal problem statements; to provide logical arguments (at an appropriate level) that the models constructed are valid; and to use the models to solve problems.

Emphasize modeling through student projects that are engaging, meaningful, and relevant to student learning and interests. Modeling is a powerful problem solving process that helps students use their skills, knowledge, and creativity to produce results and products that can benefit society. Therefore, modeling can build student confidence, introduce them to useful and powerful elements of mathematics, and provide a mechanism for communication, expression, and reasoning that is cross-cultural and cross-disciplinary.

The need to increase the emphasis on mathematical modeling in the first two years of the undergraduate program is a strong message from the Curriculum Foundations Project.

Emphasize communication skills.

- *Incorporate development of reading, writing, speaking, and listening skills into courses.*
- *Require students to explain mathematical concepts and logical arguments in words. Require them to explain the meaning — the hows and whys — of their results.*

A theme that ran through nearly every disciplinary workshop was the importance of being able to communicate mathematical and quantitative ideas. Though there are successful examples of instructors who teach writing and speaking in the mathematics classroom, there is still a need for more universal implementation of these activities. Many mathematicians view instruction in writing and oral presentation as hard, time-consuming, and foreign to their own training. However, these skills are critical to students, and faculty members in all disciplines have a responsibility to incorporate them into class instruction. Such activities can take the form of written lab assignments, technical reports, group projects, professional presentations in class, short essays on exams, and the like.

Communication skills are related to logical reasoning: if you can't explain it, you don't understand it. Therefore, these skills are essential to the development of students' mathematical knowledge in the first two collegiate years.

Emphasize balance between perspectives.

- *Continuous and discrete*
- *Linear and nonlinear*
- *Deterministic and stochastic*
- *Deductive and inductive*
- *Exact and approximate*

- *Pure and applied*
- *Local and global*
- *Quantitative and qualitative*

A broader view of mathematics needs to be communicated to students. The perspectives on the left in the above list are those that are usually covered in the undergraduate mathematics curriculum. Those on the right are, for the most part, not covered. The consensus of the Summary Curriculum Foundations Conference participants was that students should be exposed to both views, although the depth, breadth, sequence, and methods of how this should be done would depend on the nature of the local institution.

Priorities for Content, Topics, and Courses

Strive for depth over breadth. Explore locally what topics can be omitted and teach the remaining topics in more depth.

Topics can and should be eliminated to achieve a depth of conceptual understanding on a limited number of mathematical tools. Specifically, topics currently taught in some introductory courses can be omitted, and replaced with a more detailed treatment of the remaining material or with new material that better supports the desired conceptual objectives. For example, most workshops placed emphasis on derivatives and integrals as conceptual ideas and tools while downplaying the importance of intricate computational techniques. Time can be gained for conceptual underpinnings by focusing only on calculations using the most basic functions and eliminating more tedious calculations that students will not likely remember past the final exam.

Offer non-calculus-based descriptive statistics and data analysis in the first two years (either as a separate course or integrated into other courses).

The importance of data analysis for so many of the partner disciplines argues strongly for an increased presence of basic statistical training in the first two years of undergraduate mathematics. The experience provided should be primarily concerned with descriptive statistics, needing only a brief introduction to probability. The material should be motivated by a variety of examples and real data sets, including data collected by students.

Offer discrete mathematics and mathematical reasoning in the first two years. Do not have calculus as a prerequisite.

Discrete mathematics, including a serious introduction to proof techniques, is absolutely essential for students majoring in computer science. In many institutions, the instruction of discrete mathematics has been taken over by computer science departments because mathematicians do not consider it important, believe a high level of mathematical maturity is required, or do not teach it in a way that is useful to computer scientists. This is unfortunate, especially since proof techniques are an integral part of discrete mathematics.

Mathematics departments should offer at least one introductory course in discrete mathematics, preferably at the freshman level, which serves the needs of computer science and provides a bridge from introductory mathematics to more theoretical and abstract upper-division courses. Ideally, such discrete mathematics courses would require no calculus prerequisite. Computer science faculty should be heavily involved in the design and implementation of these courses.

Continue to offer calculus and linear algebra in the first two years, but make the curriculum more appropriate for the needs of the partner disciplines.

Continuous mathematics, most commonly seen in calculus courses and linear algebra, has not lost its importance for many of the partner disciplines. However, these courses have often drifted in a theoretical direction that has made them obscure, formidable, and seemingly irrelevant to other disciplines. This is unhealthy for both the partner disciplines and mathematics. Making these courses more appropriate for the partner disciplines will mean confronting the issues raised by the calculus reform movement and the less widely known, but still vital, reform efforts in linear algebra instruction. Specifically, workshop participants stressed that only the most fundamental and applicable results from calculus are needed, and that linear algebra should stress matrix algebra through eigenvalues and eigenvectors. In most workshops, representatives of the partner disciplines advocated attitudes and actions that were *exactly* those put forward during the calculus reform movement. Yet in almost every instance these colleagues had not even heard of the calculus reform movement.

Replace traditional college algebra courses with courses stressing problem solving, mathematical modeling, descriptive statistics, and applications in the appropriate technical areas. De-emphasize intricate algebraic manipulation.

College algebra courses serve two distinct student populations: the overwhelming majority for whom this is a terminal course in mathematics, and the relatively small minority for whom it is a gateway to further mathematics. Neither group is well-served by the traditional version of the college algebra course. Many of the disciplinary workshop reports recommend the reorganization of college algebra and precalculus courses to better meet the needs of various student populations. In particular, the obvious mismatch between a curriculum designed to prepare students for calculus and the reality that very few of these students subsequently enroll in calculus caused the Summary Curriculum Foundations Conference participants to recommend changes stressing problem solving, modeling, statistics, and applications.

Emphasize two- and three-dimensional topics.

The most urgent plea for an earlier introduction of multidimensional topics came from the chemists: introductory chemistry courses utilize many topics in multivariable calculus. However, colleagues from many disciplines expressed dissatisfaction with the level of student understanding of concepts—geometric and otherwise—in three dimensions. Specifically, they stressed the need for vectors in two and three dimensions, geometric and graphical reasoning, linear systems, and three-dimensional visualization skills.

Pay attention to units, scaling, and dimensional analysis.

Using units and dimensional analysis, as well as realistic scaling, is important in helping students make the transition between the concrete physical world of observations and data and the abstract mathematical world of formulas and models. Specifically, appropriate units and scale can help students to apply meaning to algebraic formulas and equations.

Instructional Techniques and Technology

Use a variety of teaching methods since different students have different learning styles. In particular, encourage the use of active learning, including

- *in-class problem solving opportunities*
- *class and group discussions*
- *collaborative group work, and*
- *out-of-class projects.*

Colleagues in partner disciplines support the need for alternatives to traditional lecture courses. Mathematics classes that are based primarily on lectures may discourage student interest and minimize opportunities for students to develop mathematical understanding. Some students learn best by demonstration, some by visual recognition, some by hands-on experience, and others by conceptualization. Therefore, it is important to create diverse learning environments within each class. In addition, active learning by students is far more likely to increase the "residue of knowledge" that students retain after a course is over.

Improve interdisciplinary cooperation.

Mathematicians should seek out projects from partner disciplines to be used in mathematics courses and increase team teaching opportunities. Disciplinary workshop participants were so excited by the possibility of increasing the use of real models in mathematics courses that many volunteered to help with the development of such models. The support from colleagues in the partner disciplines for interdisciplinary cooperation with mathematics departments was so strong that mathematicians at individual institutions should be encouraged to seek out such support locally.

Interdisciplinary cooperation can help students overcome the transfer problem from mathematics courses to partner discipline courses. Specifically, students often have difficulty seeing the relationships between problems in non-mathematics disciplines and material studied in mathematics courses. Colleagues in partner disciplines believe that exposing students in mathematics courses to discipline-specific contexts for various mathematical topics will have a positive effect on their ability to transfer knowledge between courses.

Interdisciplinary class sessions—both in mathematics courses and courses in the partner disciplines—were enthusiastically endorsed by workshop participants as a means for increasing interaction between mathematics faculty and partner discipline faculty. For example, an engineering professor presenting specific applications to a mathematics class will increase the sense of relevancy for students. Conversely, mathematics professors coming into engineering courses when specific mathematics topics are being reintroduced would reinforce earlier mathematics instruction.

Cooperative teaching arrangements between mathematics and other disciplines can be expensive. In addition to funding issues, there are often administrative barriers that need to be addressed, including questions about tenure and promotion. MAA should take a visible leadership role in helping institutions confront these issues.

Emphasize the use of appropriate technology.

Technology should be used in introductory mathematics courses to provide students with tools for solving problems. Current technologies make possible the discussion of important problems that were previously inaccessible, such as problems without analytic solutions. Technology is thus a powerful tool that should be utilized fully in the mathematics classroom. However, mathematics faculty must stress to students the importance of choosing the appropriate method of calculation (mental, paper-and-pencil, or technological) for the desired task.

Colleagues in other disciplines do not need to be convinced of the importance of technology in mathematics instruction since they know how critical technology is to their own fields. And they are well aware of the importance of choosing appropriate tools for each problem solving activity, as well as properly interpreting the results so obtained. Therefore, mathematics courses should stress intelligent and careful interpretation of results obtained from technology. Blind, unquestioning belief in the results obtained from a calculator or computer can be disastrous.

A more surprising statement from workshop participants was that spreadsheets are the technology of choice for a large number of partner disciplines. Although individual workshop reports stopped short of recommending spreadsheets as the primary technology in mathematics instruction, their widespread use is relevant to the technology choices made in mathematics courses that primarily serve other disciplines.

A related observation was the unimportance of graphing calculators; very few workshop participants reported their use in disciplinary courses. Therefore, if calculators are chosen as the technology for a mathematics course, it must be understood that this is done for pedagogical reasons, not to support uses in other disciplines.

The bottom line: mathematics faculty members need to be aware of the preferred tools of the partner disciplines—both globally and locally.

Emphasize the use of appropriate assessment.

The important relationship between assessment and student learning was discussed extensively at the workshops; i.e., how and what you assess directly affects how and what students learn. Because assessment can be difficult, time-consuming, and tedious, instructors often put less thought and effort into this aspect of course design. However, since effective assessment is critical to learning, instructors must invest in the development of a variety of assessment strategies to measure achievement of course objectives.

WYTIWYG ("What you test is what you get") was adopted at the Summary Curriculum Foundations Conference as a central message about assessment. For example, discussions focused on the need for conceptual questions on examinations as opposed to only algorithmic computations and problems that can be solved mechanically with a calculator. It underscores the importance assigned by colleagues in partner disciplines to the development of conceptual understanding.

Recommendations for Departments and the Profession

Promote professional development.

Changes in the mathematics curriculum that are advocated in this volume will not occur easily or naturally. They will require a great deal of effort, insight, and hard work, and will not happen until the mathematics community believes in the need for these changes. It is also important that MAA actively support these recommendations from the partner disciplines. However, meaningful implementation needs to be developed at the local level. To assist in this curriculum development, MAA can promote the urgency of the recommendations and widely disseminate a variety of curricular models as they become available.

Establish mechanisms for the development, review, and dissemination of effective instructional materials and techniques, including collaborative efforts between mathematicians and partner disciplines that result in innovative instructional materials.

MAA should take the lead in establishing mechanisms by which colleagues from partner disciplines can collaborate with mathematics faculty, and the products of these collaborations can be made readily available to a wider audience. Courses taught by faculty teams that include both mathematicians and partner discipline representatives should be encouraged by institutions. Several colleges and universities have implemented programs that encourage the development of interdisciplinary courses. Such programs can be studied and adapted for local implementation.

Encourage institutional assessment of programmatic changes.

The recommendations made in this volume are fundamental and far-reaching. It is therefore important that MAA, in cooperation with the mathematics community and individual institutions, engage in a sustained effort to collect and analyze information regarding the influence of these recommendations on undergraduate mathematics education. In particular,

- How do mathematics departments implement the recommendations?
- How do colleagues in other disciplines respond to the recommendations?
- How do students adjust their modes of interaction with the mathematics curriculum as a result of the recommended curricular changes?
- How do textbook writers, publishers, and educational website developers react to the recommendations?
- How is the "residue of student learning" affected at institutions that implement these recommendations?

Utilizing the Reports in this Volume

Although some differences between disciplines have been observed in the Curriculum Foundations workshops, the guiding messages are strong. These messages can be clearly understood when organized into the single Collective Vision outlined here. However, this common vision is not a substitute for the unique contributions of colleagues in the many disciplines participating in this effort. And, depending on the situation and circumstances at individual institutions, there may be a need for a greater focus on the needs of certain disciplines.

Therefore, the full reports as submitted by the participants in each of the Curriculum Foundations workshops are included here. These reports provide the details of individual discussions at these workshops as interpreted by leaders from the partner disciplines. The words are their words, the recommendations are their beliefs about their students. Individual reports can serve as a resource to guide one-on-one discussions between mathematicians and colleagues in partner disciplines. And because the reports were written by the partner disciplines, they are more likely to be received by colleagues in those disciplines as credible ideas from like-minded colleagues—not a mandate handed down by mathematicians. In this way, it is hoped that the reports will foster open, collegial, and constructive conversations between academic departments.

Finally, the reports to follow give renewed impetus for continuing efforts of the mathematics community in renewing the undergraduate curriculum. Ultimately, it is the responsibility of mathematicians to create courses and curricula that embrace the spirit of these recommendations while retaining the intellectual integrity that is so integral to the discipline of mathematics. MAA continues to provide resources for this effort including the publication of the *CUPM Curriculum Guide 2004*, addressing the undergraduate program in mathematics. However, all mathematics departments—and all mathematicians in those departments—must make the development of high quality, relevant mathematics courses at the introductory collegiate level a priority of great importance. The students who take these courses will be the leaders of tomorrow in government, business, and the arts. Valuable and meaningful mathematical experiences will contribute positively to the welfare of society and insure future support for mathematics.

ACKNOWLEDGEMENTS

The Curriculum Foundations Project was guided by a Steering Committee that provided vision and organizational inspiration for the work. Committee members were:

William Barker (Chair), Bowdoin College
Susan Ganter, Clemson University
William Haver, Virginia Commonwealth University
Deborah Hughes Hallett, University of Arizona
Harvey Keynes, University of Minnesota
William McCallum, University of Arizona
Don Small, U. S. Military Academy, West Point
Kathleen Snook, U. S. Military Academy, West Point

CRAFTY and the Steering Committee thank the colleges, universities, and research institutes that hosted the workshops, and in many cases provided the necessary financial support:

Bowdoin College
Clemson University
Grinnell College
J. Sargeant Reynolds Community College
Los Angeles Pierce College

Macalester College
Mathematical Sciences Research Institute
Michigan State University
U.S. Military Academy, West Point
University of Arizona
Virginia Commonwealth University

We particularly thank the U.S. Military Academy at West Point for hosting the Summary Curriculum Foundations Conference in November 2001. It was fitting that the Project hold its final meeting at West Point since the original idea for a series of workshops with the partner disciplines was conceived by Don Small, Professor of Mathematics at West Point.

We also gratefully acknowledge the financial support of the National Science Foundation and the American Statistical Association. The NSF generously provided the necessary funding for the two workshops in technical mathematics and the ASA generously provided the necessary funding for the workshop in statistics.

We also thank the local organizers and report editors of the individual Curriculum Foundations workshops. These people, whose names are listed at the top of each of the eighteen reports, worked tirelessly—and with very tight budgets—to bring together distinguished colleagues from both mathematics and the partner disciplines.

Finally, we would like to thank the individuals without whom none of this would have been possible: our colleagues in the partner disciplines who gave freely of their time and energy. It is their ideas that form the substance of these reports. We are excited to be continuing the dialogue with this dedicated group of individuals. The mathematics community owes a debt of gratitude to each workshop participant listed in the enclosed reports.

PUBLICATIONS RESULTING FROM THE CURRICULUM FOUNDATIONS PROJECT

Barker, W., The Curriculum Foundations Project, *FOCUS, Vol. 20, No. 8,* Mathematical Association of America, Washington, DC (Nov. 2000).

Barker, W. and S. L. Ganter, Fundamental Mathematics: Voices of the Partner Disciplines, In S. Gordon and J. Narayan (Eds.), *Rethinking Precalculus*, MAA Reports, Mathematical Association of America, Washington, DC (2004, in press).

Bigio, D. and S. L. Ganter, The Curriculum Foundations Workshop on Engineering, *FOCUS, Vol. 22, No. 9,* Mathematical Association of America, Washington, DC (Dec. 2002).

Bressoud, D., The Curriculum Foundations Workshop on Chemistry, *FOCUS, Vol. 22, No. 6,* Mathematical Association of America, Washington, DC (Aug./Sept. 2002).

Bressoud, D., What's Been Happening to Undergraduate Mathematics, *J. Chem. Educ., Vol. 78* (May 2001).

Clayton, M., Calculating a New Approach to College Math, *Christian Science Monitor*, November 9, (1999).

Craig, Norman C., Chemistry Report: MAA-CUPM Curriculum Foundations Workshop in Biology and Chemistry, *Journal of Chemical Education, Vol. 78,* 582–6 (2001)

Emery, G., The Curriculum Foundations Workshop in Physics, *FOCUS, Vol. 21, No. 3,* Mathematical Association of America, Washington, DC (March 2001).

Graham, M. D., and S. L. Ganter, The Interface Between Chemical Engineering and Mathematics: What do students really need? *Chemical Engineering Education*, American Society for Engineering Education (2001).

Oni, B. et al., Foundation Mathematics for the Study of Electrical Engineering, Institute of Electrical and Electronics Engineers (IEEE) *Transactions on Education* (submitted and under revision).

Katz, L., and S. L. Ganter, The Mathematical Needs of Civil Engineers: Renewing the undergraduate mathematics curriculum, *ASEE Prism* (2004, in press).

Moore, T., R. Peck, and A. Rossman, The Curriculum Foundations in Statistics, *FOCUS, Vol. 22, No. 9,* Mathematical Association of America, Washington, DC (Dec. 2002).

J. C. Peterson, Curriculum Foundations Workshops for Technical Programs in Two-Year Colleges, *FOCUS, Vol. 22, No. 8,* Mathematical Association of America, Washington, DC (Nov. 2002).

Salem, A., and J. Dilts, The Curriculum Foundations Workshop in Biology, *FOCUS, Vol. 23, No. 3,* Mathematical Association of America, Washington, DC (March 2003).

Small, D., and K. Snook, The Curriculum Foundations Workshop on Interdisciplinary Core Mathematics, *FOCUS, Vol. 22, No. 6,* Mathematical Association of America, Washington, DC (Aug./Sept. 2002).

Terrell, W. J., and T. F. Huff, Curriculum Foundations Workshop on Mathematics for Health and Life Sciences, *FOCUS, Vol. 22, No. 8,* Mathematical Association of America, Washington, DC (Nov. 2002).

Tucker, A., The Curriculum Foundations Workshop in Computer Science, *FOCUS, Vol. 22, No. 5,* Mathematical Association of America, Washington, DC (May/June 2002).

APPENDIX A: The Curriculum Foundations Workshops

Physics and Computer Science

Bowdoin College, October 1999

Interdisciplinary (Mathematics, Physics, Engineering)

United States Military Academy, West Point, November 1999

Engineering

Clemson University, May 2000

Health-Related Life Sciences

Virginia Commonwealth University, May 2000

Technical Mathematics (at two sites)

Los Angeles Pierce College, October 2000

J. Sargeant Reynolds Community College, October 2000

Statistics

Grinnell College, October 1999 and 2000

Business and Management

University of Arizona, October 2000

Teacher Preparation and Mathematics Education

Michigan State University, November 2000

Biology and Chemistry

Macalester College, November 2000

Mathematics Preparation for the Major

Mathematical Sciences Research Institute, February 2001

* *

The Summary Curriculum Foundations Conference

United States Military Academy, West Point, November 2001

APPENDIX B: Curriculum Foundations Workshop Questions

Understanding and Content

- *What conceptual mathematical principles must students master in the first two years?*
- *What mathematical problem solving skills must students master in the first two years?*
- *What broad mathematical topics must students master in the first two years? What priorities exist between these topics?*
- *What is the desired balance between theoretical understanding and computational skill? How is this balance achieved?*
- *What are the mathematical needs of different student populations and how can they be fulfilled?*

Technology

- *How does technology affect what mathematics should be learned in the first two years?*
- *What mathematical technology skills should students master in the first two years?*
- *What different mathematical technology skills are required of different student populations?*

Instructional Interconnections

- *What impact does mathematics education reform have on instruction in your discipline?*
- *How should education reform in your discipline affect mathematics instruction?*
- *How can dialogue on educational issues between your discipline and mathematics best be maintained?*

Instructional Techniques

- *What are the effects of different instructional methods in mathematics on students in your discipline?*
- *What instructional methods best develop the mathematical comprehension needed for your discipline?*
- *What guidance does educational research provide concerning mathematical training in your discipline?*

Biology

CRAFTY Curriculum Foundations Project
Macalester College, November 2–5, 2000

Judy Dilts and Anita Salem, Report Editors
David Bressoud, Workshop Organizer

Summary

It is generally agreed that research in biology has become more quantitatively oriented than in the past. At the same time, it is also recognized that the quantitative needs of undergraduate students enrolled in biology courses are diverse and depend largely upon the student audience (e.g., majors versus non-majors) and the variety of disciplinary tracks, ranging from molecular biology to ecology, that students choose to explore. In an already crowded biology curriculum, the panelists agreed that the issue of increasing quantitative emphasis would call for innovative solutions. They suggested solutions ranging from the creation of mathematical courses designed specifically for biology majors to the creation of mathematical modules that could be incorporated into existing biology courses.

To build and require more quantitatively oriented biology courses would be a major, but important, undertaking and would necessitate increased cooperation among biologists and mathematicians. The proposed actions of the MAA in assisting their client colleagues with possible changes and emphasis in the mathematics curriculum could serve not only to increase the quantitative literacy of biologists, but also act as a catalyst for needed changes in the undergraduate biology curriculum. Some common themes that emerged during the workshop were:

1. New areas of biological investigation together with advances in technology have resulted in an increase in quantification of biological theories and models.
2. The collection and analysis of data that is central to biological investigations inevitably leads to the use of mathematics.
3. Mathematics provides a language for the development and expression of biological concepts and theories. It allows biologists to summarize data, to describe it in logical terms, to draw inferences and to make predictions.
4. Statistics, modeling and graphical representation should take priority over calculus.
5. The teaching of mathematics and statistics should use motivating examples that draw on problems or data taken from biology.
6. Creating and analyzing computer simulations of biological systems provides a link between biological understanding and mathematical theory.

Narrative

Introduction and Background

The proposition that is being addressed in this report is how undergraduate mathematics education can better serve students majoring in biology. The comments that follow are predicated on these assumptions:

1. Many undergraduate students taking introductory mathematics courses have educational destinations other than mathematics.
2. The sciences are increasingly seeing students who are quantitatively ill-prepared.
3. The biological sciences represent the largest science client of mathematics educators.
4. Students majoring in biology typically complete only one or two semesters of mathematics.
5. The current mathematics curriculum for biology majors does not provide biology students with appropriate quantitative skills.
6. The field of biology is becoming much more quantitative which will necessitate a change in the mathematics curriculum for biology majors.

One particular challenge facing biology educators is the range of mathematical backgrounds of professors of biology. Many, if not most, biology educators have completed only calculus and one course in statistics. The limited mathematical background of most biologists is clearly reflected in the correspondingly limited quantitative components of both biology textbooks and curricula. As we begin to expand the quantitative backgrounds of our students we will also have to provide opportunities for the biology faculty to increase their own facility with mathematics.

Understanding and Content

Surveys of quantitative skills needed for biologists frequently include college algebra, introductory calculus and statistics. Among these three areas of mathematics, statistics is the most commonly mentioned and the most extensively used. Other content areas that are mentioned include mathematical modeling, discrete mathematics and matrix algebra.

1. **College Algebra*:*** Biology students need to understand the meaning and use of variables, parameters, functions and relations. They need to know how to formulate linear, exponential and logarithmic functions from data or from general principles. They must also understand the basic periodic nature of the sine and cosine functions. It is fundamentally important that students are familiar with the graphical representation of data in a variety of formats (histograms, scatter plots, pie charts, log-log and semi-log graphs.)
2. **Introductory Calculus:** The topics from introductory calculus that were mentioned at the workshop included integration for the purpose of calculating areas and average value, rates of change, optimization, and gradients for the purpose of understanding contour maps.
3. **Statistics:** It is here where the list of necessary topics is the longest and encompasses descriptive statistics, conditional probability, regression analysis, multivariate statistics, probability distributions, simulations, significance and error analysis.
4. **Discrete Mathematics and Matrix Algebra:** The topics most frequently mentioned were qualitative graphs (trees, networks, flowcharts, digraphs), matrices (Leslie, Markov chains), and discrete time difference equations. Other topics included equilibria, stability and counting techniques.

Technology

The pervasive presence of computers together with their ever-increasing computational power encourages biologists to apply statistical methods to analyze data that is collected in the laboratory or the field. One important software application used by biologists is the spreadsheet. Increasingly, spreadsheet applications contain sophisticated statistical tools sufficient for use with undergraduate biology majors. The panelists were unanimous in their observation that *the graphing calculator is not the tool of choice* for biology students. Technological tools must be capable of producing graphs that can be incorporated into printed and presentation documents. They must allow students to apply modeling techniques to large data sets and they must also support simulation of models that are stochastic, discrete or continuous.

Implementation

There is a variety of ways to implement curricula containing the recommended mathematical topics. It should be noted that responsibility for student competence with mathematics should not rest solely in the hands of the mathematics faculty. The biology faculty must incorporate the use of mathematics into their courses in order to reinforce and verify the importance of mathematics to their students. To this end, the quantitative courses must be taken early so that the topics introduced can be used in subsequent courses in biology. Collaborative efforts to design and deliver the quantitative courses should be encouraged. Some possible curricular options considered by the panel include:

1. Mathematical requirements may be completed in a one-semester course. The panel does *not recommend* this option, however if only one course in mathematics is required, they suggest that the emphasis of this course be on statistics.
2. Mathematical requirements may be completed in a two-semester sequence. In this scenario, the panel recommends that one course focus on topics in statistics and the other course include topics from calculus and mathematical modeling. A yearlong course integrating these three topics would be preferable.
3. Mathematical requirements may be completed in a three-semester sequence. This option would include a course in statistics and a course integrating topics from calculus and mathematical modeling. The third course might come later in the curriculum as a project-based course oriented toward modeling applications and could also include topics from matrix algebra and discrete mathematics.
4. Mathematical requirements may be completed by an integrated approach. An interdisciplinary approach could be used to embed mathematical modules into biology courses. This approach has the advantage of teaching the mathematical concepts in the context of biological applications. The challenges to this approach are the usual difficulties that arise in the design and delivery of interdisciplinary courses.

WORKSHOP PARTICIPANTS

Biology Participants

Robert Blystone, Professor of Biology, Trinity University
David Campbell, Professor of Biology, Grinnell College
Mark Davis, Professor of Biology, Macalester College
Judy Dilts, Professor of Biology, William Jewell College
Louis Gross, Professor of Biology and Mathematics, University of Tennessee
Dan Hornbach, DeWitt Wallace Professor of Biology, Macalester College
John Jungck, Professor of Biology, Beloit College
Gary Reiness, Professor of Biology, Lewis and Clark College
Jan Serie, O.T. Walter Professor of Biology, Macalester College
Dwight Underhill, Henry Savage, Jr. Professor of Public Health Research, U. of South Carolina

Mathematics Participants

Fred Adler, Associate Professor of Biology and Mathematics, University of Utah
David Bressoud, DeWitt Wallace Professor of Mathematics, Macalester College
Daniel Kaplan, Associate Professor of Mathematics, Macalester College
Eric Marland, Assistant Professor of Mathematical Sciences, Appalachian State University
Stephen Merrill, Professor of Mathematics, Marquette University
Weiwen Miao, Assistant Professor of Mathematics, Macalester College
Richard Molnar, Associate Professor of Mathematics, Macalester College
Claudia Neuhauser, Professor of Ecology, Evolution and Behavior, University of Minnesota
Anita Salem, Professor of Mathematics, Rockhurst College
Daniel Tranchina, Associate Professor of Mathematics and Biology, Courant Institute, New York University.

ACKNOWLEDGEMENTS

The members of the panel would like to thank their colleagues in the Mathematical Association of America for providing the opportunity to consider and make recommendations on a topic of importance: William Barker, the chair of CRAFTY, and David Bressoud for coordinating the workshop at Macalester College. We are grateful for support of the workshop by a grant from the National Science Foundation, an institutional grant from the Howard Hughes Foundation, and Macalester College.

Business and Management

CRAFTY Curriculum Foundations Project
University of Arizona, October 28–29, 2000

Chris Lamoureux, Report Editor
Lee Beach and Deborah Hughes Hallett, Workshop Organizers

Summary

Mathematics Departments can help prepare business students by stressing problem solving using business applications, conceptual understanding, quantitative reasoning and communication skills. These aspects should not be sacrificed to breadth of coverage.

Problem solving includes developing and applying appropriate abstract models and an understanding that many quantitative problems are ambiguous and uncertain. Business faculty would like their students to be comfortable taking a problem and casting it in mathematical terms. Conceptual understanding can be fostered by motivating the mathematics with the "whys"– not just the "hows." Quantitative reasoning includes becoming familiar and comfortable with the language of mathematics and the application of mathematical reasoning to quantitative problems. Quantitative literacy can also be developed through the use of the technological tools of business, such as spreadsheets, as distinct from software whose primary applications are in science. Communication skills can be developed by having students work in teams and communicate solutions through oral and written reports held to the same professional standards they will meet in the business world.

Business students, although able, are often math phobic. Courses should strive to lessen math phobia, enable students to be more comfortable with mathematics, and help students appreciate the relevance of mathematics.

Narrative

Understanding and Content

In general, business faculty are less concerned with specific course content than with developing quantitative literacy and analytical ability in our students. Upon completion of a business mathematics curriculum the students should be comfortable with using mathematics as a tool to communicate analytical concepts. A measure of the curriculum's success is the students' comfort level when exposed to a new formula in a business class.

When in doubt, mathematics faculty should cover *less* material—and treat the material covered with respect—imparting to the students a sense of the importance of mathematics as a necessary part in the development of successful business people.

Business decisions are most commonly made under conditions of uncertainty and risk. Inferences must be drawn from data and information that are incomplete, inconclusive, and most likely imprecise. Wherever possible, math courses should attempt to illustrate this ambiguity and provide guidance in dealing with such uncertainty and variation. Sensitivity analysis can be used to demonstrate how changes in assumptions, variation in data, and the influence of contextual variables will affect outcomes. This approach will facilitate a more in-depth understanding of the interrelationships and interdependencies embedded in real world situations.

In order to achieve the desired outcomes, the business faculty recommends that the curriculum include:

1. Realistic business problems. We do not expect mathematics faculty to develop problems on their own. We envision a partnership, in which business faculty contribute.
2. Solutions that make use of business technology, such as spreadsheets.
3. Real (or realistic) data sets.
4. Problem motivated modeling.
5. Development of students' abilities to express ideas symbolically.
6. Sensitivity analysis.

Algebra is a basic prerequisite to study in business, and should either be validated by a placement test or taken before the business mathematics curriculum. Students should be able to solve simultaneous equations, understand the concept of a function and functional relationships, understand the use of common functions in modeling business concepts, construct and understand graphs, and use abstraction to build simple models.

Calculus in the business mathematics curriculum should emphasize the basic concepts and how they apply to business problems, with more attention to numerical methods and less to techniques of symbolic differentiation and integration. The business calculus curriculum should include an introduction to rates of change, and the dynamic nature of real world systems, constrained optimization, and interpretations of area under a graph.

Statistics and probability in the business mathematics curriculum should include measures of central tendency and dispersion, probabilities (including conditional probabilities and decision tree analysis), discrete and continuous probability distributions, and hypothesis testing. Students should examine, summarize, analyze, graph and interpret real data sets used in business.

In virtually all business schools, there is a requirement for an additional statistics course beyond the introductory course in basic probability, usually taught by business faculty. Regardless of who teaches this course, it should also be integrated seamlessly into the business mathematics curriculum.

Technology

Technology has several roles to play in the business mathematics curriculum. First, it provides tools that students will encounter in the work place. Second, it enhances the effectiveness and efficiency of the learning process. Third, it can help to deepen and maintain student interest.

Technology has revolutionized the way in which business is practiced. Loan payments, the value of a bond, the effect of a change in sales on net income or the price of an option are computed with a financial calculator or a spreadsheet. Debits and credits, and journal entries no longer dominate the teaching of accounting classes. The algebraic manipulation necessary to compute present and future value no longer dominates finance courses. The business leaders of tomorrow, and therefore the business students of today, need to understand the conceptual basis of algebra, calculus and statistics. They must be able to interpret and use the results of calculations. The accountant has changed from a bean counter who reports what has already occurred into a business planner who assists charting the future course of the company. The finance executive has to be able to quickly evaluate several competing alternatives with risk and expected returns. For business executives to be successful, they need proficiency in the technology that produces the data they need, understanding of the algebra, calculus and statistics underlying these data, and knowledge of how sensitive the results are to changes in the input data.

Spreadsheets are very useful in charting data, conducting exploratory data analysis, developing models, demonstrating impact of changes on the inputs on the outputs, carrying out parametric analysis, and in more sophisticated applications such as optimization and simulation. Their potential should be fully exploited in the classroom, and the classroom use should be complemented with hands-on experience for the students in computer labs. The labs can be self-directed and rely on tutorials available on the web or prepared for the class, or they can take the form of instructional labs or workshops supported by teaching assistants, depending on the needs and resources of the institution. Students could repeat the in-class exercises or work on assignments and projects in the labs.

Specific technology goals for the curriculum include:

1. Comfort with the use of technology as an analytical tool.
2. Integrated spreadsheet experiences starting with mathematics and continuing in subsequent courses.
3. Using technology to answer "what-if" questions.
4. Encouraging students to experiment and try alternative approaches to a problem.
5. The use of realistic data.
6. Enhancing students' ability to design an experiment and find data, e.g., from the web.
7. Enhancing students' abilities to explain calculations and prepare a clear, effective presentation.
8. Enhance students' abilities to display results graphically.

Instructional Techniques

Useful pedagogical techniques include questions to make students think about a problem or a concept, asking students to discuss questions amongst themselves, and inviting students to share their proposed solution to a problem with the class and asking the rest of the class to critique it. A very effective active learning method is real-time problem solving or model building using student input in class. Starting with a blank spreadsheet on the screen and filling it out using student suggestions and instructions gives students a sense of control over the lecture and draws them in. Another potentially useful method for drawing students into the lecture is to start the lecture with a real-world (or realistic) business problem. If students are convinced that the problem is worthy of their attention, and that they do not know how to solve it, they are much more likely to pay attention and to retain what they learn. It is important to get the buy-in at the beginning.

Many business courses attempt to develop and reinforce skills that are highly valued by employers such as group work and communication skills. We believe these skills can and should be emphasized in mathematics courses as well. Students could work on group projects or assignments, and be asked to prepare written reports or oral presentations to communicate the results of their analysis. The ability to work effectively in groups and the ability to explain quantitative concepts and results in plain English are highly valued skills in business schools, and emphasis of these skills in first-year courses would achieve a more seamless transition to business school for students.

Business students are anticipating entry into the business world. Faculty should expect that the materials that students are asked to prepare are analogous to materials that they will hand to their bosses in the business world.

How Do We Get There?

Critical to the accomplishment of goals above are cooperation and communication between the business and the mathematics faculties. The willingness to change will be built upon personal relationships. Collaborative identification of clear objectives, and creation of the corresponding pedagogy, will build the foundation for successful implementation. The probability of lasting change will depend on the monitoring and assessment of relevant performance measures.

Specific suggestions for moving towards the goal of improved mathematical competence of undergraduate business students include:

1. Improved communication and collaboration between business and mathematics faculty.
2. Treating the selection of instructors for business mathematics courses as equally important as the selection of instructors for other mathematics courses.
3. Development of a seamless course structure, where students and faculty see the mathematics curriculum as integrated with the business curriculum.
4. A shift from business mathematics viewed as a method of weeding out students to business mathematics with the purpose of adding value.
5. Improved availability of business faculty for providing motivating examples and background for business mathematics.
6. Encouraging active dialogue between MAA and AACSB, and between mathematics and business, at regional and national business conferences.
7. Recognition that each school is unique, and that no single business mathematics curriculum will work everywhere.

WORKSHOP PARTICIPANTS

Business Participants

Urton Anderson, Associate Dean for Undergraduate Programs, McCombs School of Business, The University of Texas at Austin

Lee Beach, Vice Dean, College of Business and Public Administration, University of Arizona

George Bohlander, Professor of Management and Policy, School of Business, Arizona State University

George Carter, Professor of Economics and International Business, College of Business Administration, University of Southern Mississippi

Al Croker, Chair of Statistics and CIS Department, Zicklin School of Business, Baruch College

Edward Davis, Dean, School of Business, Clark Atlanta University

Erhan Erkut, Hamilton Professor of Business, University of Alberta

Paul Fischer, Jerry Leer Professor in the School of Business Administration, University of Wisconsin-Milwaukee

Kent Foster, Chair, Computer Sciences and Quantitative Methods, College of Business Administration, Winthrop University

Sharon Garrison, Senior Lecturer, Finance, University of Arizona

Virginia Gibson, Associate Professor of Management Information Systems, University of Maine

Jim Haskins, Assistant Professor, Division of Accounting and Finance, University of North Dakota

Joseph Horton, Dean, College of Business Administration, University of Central Arkansas

Max Jerrell, Professor of Economics, College of Business Administration, Northern Arizona University

George A. Johnson, Professor of Management, Idaho State University

Vijay Kannan, Assistant Professor of Operations Management, Utah State University

Jerome Keating, Professor of Management and Marketing, University of Texas at San Antonio

Nelson Lacey, Professor of Finance & Operations Management, Isenberg School of Management, University of Massachusetts

Chris Lamoureux, Head, Finance Department, University of Arizona

Jim Lukawitz, Faculty Director Undergraduate Programs, Fogelman College of Business and Economics, The University of Memphis

James Marsh, Professor and Chair, Department of Financial Economics, University of Hawaii at Manoa

Constance McLaren, Professor of Quantitative Methods, School of Business, Indiana State University

Bruce McLaren, Professor of Management and Information Systems, Chair, Organization Department, School of Business, Indiana State University

Amit Mitra, Associate Dean, College of Business, Auburn University

Cyrus K. Motlagh, Associate Professor of Statistics and Management Sciences, College of Business Administration, University of Detroit Mercy

Scott O'Leary-Kelly, Assistant Professor, CISQA Department, Sam M. Walton College of Business, University of Arkansas

Ted Oleson, Economics Faculty, College of Business Administration, University of Nevada, Reno

Dennis Placone, Professor of Economics, College of Business and Public Affairs, Clemson University

Bruce Raymond, Associate Dean, College of Business, Montana State University

Violet Rogers, Professor of Accounting and Associate Dean, College of Business, Stephen F. Austin State University

Pam Slaten, Senior Lecturer, Assistant Department Head, Management Information Systems, University of Arizona

Timothy Sullivan, Associate Professor of Economics, College of Business and Economics, Towson University

Charles O. Tustin, Professor of Management, School of Business Administration, Fort Lewis College

Terry Weisenberger, Associate Dean, Robins School of Business, University of Richmond

Andrew Welki, Assistant Professor of Economics, Boler School of Business, John Carroll University

Sharon D. White, Assistant Professor, School of Business and Economics, North Carolina A&T State University

Mathematics Participants

William Barker, Professor of Mathematics, Bowdoin College

Deborah Hughes Hallett, Professor of Mathematics, University of Arizona

Jim Lightbourne, Division of Undergraduate Education, National Science Foundation

William McCallum, Professor of Mathematics, University of Arizona

Richard B. Thompson, Associate Professor of Mathematics, University of Arizona

James Washenberger, Assistant Professor of Mathematics, Department of Mathematics, Virginia Polytechnic Institute & State University

ACKNOWLEDGEMENTS

The conference was funded by the Eller College of Business and Public Administration at the University of Arizona and by National Science Foundation CCLI grant 9972350. The organizers would like to thank Mark Zupan, Dean of the College of Business and Public Administration, for his support.

APPENDIX: Research on Business Education

Business faculty members conduct substantial research on pedagogical issues. Periodic reviews of business education journals by CRAFTY would keep MAA current on business pedagogical issues and provide CRAFTY with mathematics initiatives and excellent business examples. The business education journals include: *Financial Practices and Education, Journal of Financial Education, Journal of Economic Education, Accounting Educators Journal, Issues in Accounting Education, Journal of Management Education, Journal of Marketing Education, INFORMS Transactions on Education,* and *Journal of Education for Business*. Research on business education is taking on an increasing importance as AACSB assessment expectations rise. In reaction, pedagogy is rising in importance and new pedagogical journals are under consideration; for instance, the Decision Sciences Institute is planning an education journal for Information Sciences. The business environment is rich with a variety of activities. There are a number of business pedagogical seminars sponsored by foundations such as the Lilly Foundation. Book publishers often conduct seminars on learning theory and the learning impact of their products, and professional associations such as the American Marketing Association, Academy of Management, Decision Sciences Institute, Financial Management Association, American Accounting Association, and the American Economic Association routinely sponsor teaching and learning sessions.

AACSB constantly monitors changes in the business environment and the corresponding changes in business education. Through networking with the AACSB staff, CRAFTY can stay apprised of the latest expectations in business content and technology. Then, real-time changes in the content-pedagogy-technology triad can be promulgated to departments of mathematics. CRAFTY reviews of the two-year mathematics curriculum might thereby become semiannual or annual rather than decennial reviews. An added benefit of this reduced review period is that mathematics faculty members become increasingly knowledgeable business school partners.

Chemistry

CRAFTY Curriculum Foundations Project
Macalester College, November 2–5, 2000

Norman C. Craig, Report Editor
David Bressoud, Workshop Organizer

Summary

Six themes for the mathematical preparation of students in the chemical sciences emerged from the chemistry workshop. These themes are (1) emphasizing multivariable relationships from the outset; (2) illustrating numerical methods applied to practical problems; (3) recognizing that chemists and biochemists depend heavily on visualizations, which are often three-dimensional; (4) paying attention to scale, units and estimation; (5) knowing that skill in mathematical reasoning is widely applicable; and (6) including practice in analyzing large data sets. Theme 5 is an especially important contribution. The mathematicians play an essential role in helping chemistry students learn how to think in logical and formal ways that are essential in the chemical sciences. Chemistry students also must learn how "to listen to the equations."

The teaching of mathematical methods for chemistry is a responsibility shared by chemists and mathematicians. The role of chemists in teaching mathematical methods increases with the level of the course work in most areas of chemistry. To be concrete about this sharing of responsibility between chemists and mathematicians, the members of the workshop developed an explicit classification presented in tabular form. Four categories for the teaching of mathematics are distinguished. They are (1) mathematical material learned in secondary school, which chemists reinforce; (2) mathematical material chemists expect mathematicians to develop; (3) mathematical material for which chemists have responsibility at the general chemistry level; and (4) mathematical material for which the chemists have responsibility in advanced courses.

No pattern in the use of powerful software tools for mathematics exists in chemistry instruction at present. Chemists use *Mathematica*, *Matlab*, and *Mathcad* to variable extents. Thus, chemists do not recommend standardizing on one of these. However, chemists make extensive use of spreadsheets. More use of spreadsheets in calculus courses would anticipate their use in chemistry. In addition, chemists are more dependent on computers than on graphing calculators.

Chemists are as concerned about the writing skills of students as are the mathematicians. Thus, chemists welcome the new emphasis on having students in mathematics express concepts in clear writing as well as symbolically. For instruction in mathematics to have greater relevance to other disciplines, problems are needed that are of disciplinary significance but expressed in terms understandable to mathematicians and their students. Chemists recommend that mathematicians actively solicit such problems and make them widely available.

Narrative

Introduction and Background

Mathematics in the Chemistry Curriculum. The range of mathematics needed for chemistry is enormous. The range extends from the relatively modest levels used by synthetic organic chemists through greater use by inorganic chemists and biochemists to extensive use by physical chemists, computational chemists and theoretical chemists. For theoretical and computational chemists the level and uses of mathematics are without limit. For analytical chemists, statistics is much more important than in other areas of chemistry. For all of the subfields within chemistry there is dependence on thinking quantitatively and on using spatial models of molecular systems. All chemists agree that undergraduates should have a good foundation in mathematics.

All chemistry and biochemistry students and students in other variations on the chemistry major take at least two semesters of calculus. Often, three semesters of mathematics are required for the B.S. major. In addition, many of the B.S. students take a semester of linear algebra or differential equations. In designing the mathematics curriculum, it is important to recognize that the first two semesters of calculus are common to all versions of chemistry majors.

It is exceptional for chemistry students to take a course in statistics in a mathematics department. Nonetheless, considerable instruction in statistics and use of statistical inference occurs in analytical chemistry courses and to a lesser extent in physical chemistry courses.

Lower-division courses, including first-year courses and organic chemistry courses, serve a wide range of students in various majors in the sciences, including biology, neuroscience, earth sciences and engineering. Few students take these courses as part of a broad education, although some students do so before switching to a major outside the sciences. Of the students who take courses in general chemistry, a large fraction intend to major in the biological sciences or to complete the requirements for medical school. Another substantial fraction in many universities and at some colleges are in engineering.

For the first-year courses, chemists do not assume that all students have taken or are taking calculus, although special sections are often taught for the mathematically inclined. Those students who will major in chemistry and biochemistry are a modest fraction (5 – 10%) of the students enrolled in first-year chemistry and organic chemistry. Despite the range of the audience for lower-division courses, it is appropriate to regard all these students as interested in growing in scientific and mathematical proficiency. For the purpose of the workshop, however, the chemistry group followed the charge presented and focused on the preparation of chemistry and biochemistry majors.

Several challenges in teaching chemistry are related to mathematical usage. One such problem is making connections between the real world of tangible chemical material and the abstractions that are used to understand these materials and their transformations. An important aspect of these abstractions lies in relating the microscopic (nanoscale) world of molecules to the macroscopic world of real chemical material. The models that describe the nanoscale world are often mathematical, and the bridges between the nanoscale world and the macroscopic world are generally crossed with mathematical reasoning. The mathematical description of chemical material is typically multivariate, a reality that does not correlate well with the emphasis on single variable functions in the typical first two semesters of instruction in calculus. In addition, chemistry students would be better prepared if they encountered a variety of variables other than the standard x, y, z set.

Chemists are aware of the calculus reform movement, but they are generally unaware of its outcomes. At the opening session of the workshop David Bressoud provided a helpful summary of the principal outcomes of calculus reform.

Perspectives of the Workshop. The group of chemists assembled for the workshop at Macalester College was weighted toward physical chemists. In addition, some were analytical or inorganic chemists by training. No organic chemists or biochemists were in the group. Two of the participants (Craig and Engstrom) were members of the Committee on Professional Training (CPT) of the American Chemical Society (ACS). They were charged with linking the deliberations in the Mathematics Workshop to the work of the CPT, which develops guidelines for undergraduate curricula in chemistry and its subfields and reviews compliance of chemistry departments with these guidelines.

The chemistry group focused on questions in the Understanding and Content cluster and in the Technology cluster of the instructions from the MAA. The chemists had some difficulty in making a sharp distinction between conceptual principles and problem-solving skills, which were distinguished under the heading of Understanding and Content. That blurring is evident in some of the specific recommendations from the chemistry study group.

Six Themes. Six themes for the mathematical preparation of chemistry students emerged from the extensive discussion of the first two clusters of questions. Mathematicians are asked to keep these six themes in mind as courses in mathematics are redesigned. Reinforcement in mathematics courses of student learning in the areas of the six themes makes the teaching of chemistry more effective. A consideration of each of the six themes follows.

1. *Multivariable Relationships*. Almost all problems in chemistry from the lowly ideal gas law to the most sophisticated applications of quantum mechanics and statistical mechanics are multivariate. Thus, it is desirable that calculus courses address multivariable problems from the outset.
2. *Numerical methods*. Numerical methods are used in a host of practical calculations in chemistry, most enabled by the use of computers.
3. *Visualization*. Chemistry is highly visual. Synthetic chemistry, which involves understanding the properties and transformations of small and large molecular assemblies, depends on practitioners being able to visualize structures and atomic and molecular orbitals in three dimensions. Understanding the consequences of quantum mechanics for chemical bonding and appreciating graphical representations, often multidimensional, depend on sophisticated visualizations.
4. *Scale and estimation*. The stretch from the nanoscale world of atoms and molecules to tangible material is of the order of Avogadro's number, which is about 10^{24}. Microscale chemistry is 10^6 or so smaller than tangible size (micrograms versus grams). Laser pulses that interrogate intimate changes in molecular species during chemical reactions may be only 10^{-15} s in duration. Other processes of interest occur on time scales up to the age of the earth (10^{17} s). Thus, distinctions in scale, along with an intuitive feeling for the different values along the scales of size, are of central importance in chemistry. Back-of-the-envelope calculations done to order-of-magnitude accuracy are sufficient in some cases and essential for checking the reasonableness of more detailed calculations. The careful use of units is an aid to approximations as well as to exact calculations.
5. *Mathematical reasoning*. Facility at mathematical reasoning permeates most of chemistry. Students must be able to follow and apply algebraic arguments, that is, "listen to the equations," if they are to understand the relationships between various mathematical expressions, to adapt these expressions to particular applications, and to see that most specific mathematical expressions can be recovered from a few fundamental relationships in a few steps. Logical, organized thinking and abstract reasoning are skills developed in mathematics courses that are essential for chemistry. At the physical chemistry level students must be able to follow logical reasoning and proofs, which is enabled by previous experience in mathematics courses. Careful use of notation also needs to be reinforced.
6. *Data analysis*. Data analysis is a widespread activity in chemistry that depends on the application of mathematical methods. These methods include statistics and curve fitting.

Answers to the Mathematicians' Questions. Because the principal goal of the workshop was to provide advice for the planning and teaching of the mathematics curriculum, providing a full account of the role of mathematics in the teaching of chemistry might be regarded as unnecessary. It seemed to the chemists, however, that attempting to isolate the part of instruction in mathematics for which the mathematicians alone have principal responsibility would be misleading. The importance and scope of mathematics in chemistry would not be fully apparent. Opportunities for linkages between chemistry and mathematics would be lost. Furthermore, chemists are unaccustomed to isolating their use and teaching of mathematics from what students have learned from mathematicians. Thus, the chemists provided a comprehensive survey of the role of mathematics in chemistry and of how instruction in mathematics should be delivered as part of the whole.

In discussions with the mathematicians the chemists confirmed their suspicion that geometry has been largely squeezed out of the secondary school curriculum. Little background in geometry helps explain why chemistry students have growing difficulty with the spatial relationships that are at the heart of much chemical thinking. The disappearance of plane geometry also removes a significant early exposure to formal proofs.

Understanding and Content

In their discussion of concepts and skills, the chemists identified several categories for specifying the locus of responsibility for instruction in mathematics. Specific conceptual principles are listed in the table in Appendix A, and specific skill areas are listed in the table in Apendix B.

The first category, designated **1** in the tables in the appendices, is conceptual material or skills that chemists expect students to bring to the first-year chemistry course from their preparation in secondary school. Chemists recognize that students will have to be reminded of many of these concepts or skills and have them reinforced and extended in the context of the chemistry course. Such teaching is the responsibility of the chemistry faculty.

The second category, designated **2** in the appendices, is conceptual material and skills that chemists expect the mathematics faculty to develop.

The third category, designated **3**, is conceptual material and skills in mathematics for which the chemists have principal responsibility. Some of this material is of a more general nature and would be developed in the lower division courses at the same time students are learning calculus. This type of instruction is designated **3G** in the appendices, where **G** stands for general. Other material is of an advanced nature and would be developed in advanced chemistry courses after students have taken the expected mathematics sequence in college. In the appendices, this type of instruction is designated **3A** , where **A** stands for advanced. Such teaching of mathematical concepts in the context of chemistry is especially effective. Revisiting mathematical methods and seeing them applied in a different context also reinforces learning.

In the classification scheme in Appendix A and Appendix B, just described, it is category **2** that is of direct importance to the mathematicians as they plan their courses. Recognizing that the list of desirable concepts for mathematics instruction is long, the concepts are prioritized in two groups. Those that are given the highest priority are in **boldface type**. The others are of second priority.

In the summing-up session with biologists, chemists and mathematicians present, the chemists were asked how the use of mathematics relates to the typical education in chemistry for students in the biological sciences. The classification scheme in the appendices provides a direct answer. The experience for students in the biological sciences consists of categories **1** and **3G**.

Technology

Technology makes it possible to address old questions more quantitatively and more realistically than was possible in the past. The complexities of real chemical material can be approached more fully. Examples of old questions are chemical equilibrium, chemical bonding, reaction mechanisms, and interpretation of spectral data. In general, solving these problems depends on multivariate analysis and numerical methods. Use of computers is assumed.

The usefulness of graphing calculators does not rank high with chemists. Chemists worry that the indiscriminate use of graphing calculators in high school mathematics may interfere with students' learning the basic concepts. Graphing calculators do provide an effective means of exploring quantitative relationships once students have mastered the fundamentals of those relationships. Chemists use the software tools *Mathematica*, *Matlab*, or *Mathcad*; *Mathcad* is the most popular. However, no pattern in the use of these materials in chemistry justifies a recommendation about standardizing use in mathematics courses. Of widespread use in chemistry teaching and research are spreadsheets, including the graphing and statistical analysis features. Spreadsheets can be used to show graphically how functions respond to changes in parameters and to show how approximations evolve. Chemists would welcome the use of spreadsheets to teach calculus. Chemists are more likely to use computers than calculators for most applications.

Uses of Technology

1. Multivariate modeling and visualization.
2. Iterative solutions.
3. Access to and use of databases such as those of the Cambridge Crystallographic Data Centre, National Institute of Standards and Technology files for smaller molecules, Beilstein, Chemical Abstracts Services, and the Protein Data Bank.
4. Data collection—high speed and extensive.
5. Data analysis such as in Fourier transform nuclear magnetic resonance and most other forms of modern spectroscopy.
6. Experimental design.
7. Pattern recognition.
8. Combinatorial chemistry in which numerous variations on a chemical reaction are run in parallel and then the efficacy of the various products is tested. Such methods are now widely applied for drug discovery, for developing new light emitting diodes and the like. Robotics and computer handling of the plethora of data are essential features of such work.
9. Facilitating interdisciplinary investigations. Internet collaborations.

Instructional Techniques

Today's mathematicians and chemists agree on the value of having students write to learn mathematics and chemistry. A laudable goal of the calculus reform movement is to have students write to foster critical thinking skills. Chemistry students, in particular, need to be confident about mathematics as an active language. Chemists have become more systematic in having their students write significant reports and critiquing these reports in supportive ways. The chemists applaud similar efforts within the mathematics community.

Instructional Interconnections

The chemists agreed that communication has been inadequate between mathematicians and chemists regarding the curriculum of mutual interest. They applauded working toward reform in communication between the disciplines as well as reform within the disciplines.

A concrete proposal for strengthening the linkage between mathematicians and chemists is to have chemists provide a set of representative problems in which various mathematical methods are crucial. These chemical problems would be expressed in language understandable to mathematicians. These problems should also be ones that could be used to teach students in mathematics courses. The initiative for starting such a process lies with the mathematicians.

WORKSHOP PARTICIPANTS

Mathematics Participants

Ronald L. Christensen, Physical Chemistry, Bowdoin College
Norman C. Craig , Physical Chemistry, Oberlin College
Glenn A. Crosby, Physical/Inorganic Chemistry, Washington State University
William F. Coleman, Physical/Inorganic Chemistry, Wellesley College
Royce C. Engstrom, Analytical Chemistry, University of South Dakota
Peter C. Jurs, Analytical/Computational Chemistry, Pennsylvania State University
Joseph J. Lagowski, Inorganic Chemistry, University of Texas Austin
A. Truman Schwartz, Physical Chemistry, Macalester College
Theresa Julia Zielinski, Physical Chemistry, Monmouth University

Mathematics Participants

David Bressoud, Mathematics, Macalester College
Thomas Halverson, Mathematics, Macalester College
Roger Howe, Mathematics, Yale University

ACKNOWLEDGEMENTS

The participants in the chemistry workshop thank Macalester College for the fine arrangements and the warm welcome we enjoyed during our work on this project. David Bressoud played a crucial role in setting up the workshop, a gathering that provided opportunities for chemists and biologists to discuss the important matter of undergraduate instruction in mathematics with mathematicians. We also thank the CUPM of the MAA for having the wisdom to consult with faculty members in other disciplines whose students depend on instruction in mathematics. We are grateful for support of the workshop at Macalester College by a grant from the NSF, by an institutional grant from the Howard Hughes Foundation, and by Macalester College.

APPENDIX A. Understanding and Content, Conceptual Principles Classification Scheme

Who Teaches Material*

		1	2	3G	3A
1.	Basic mathematics. Algebra; scientific notation; graphing; ratios; percent; shapes, simple geometry, Platonic solids; solution to the quadratic equation; functions and their graphical expression: exponential, logarithmic (base *e* and base 10), polynomial, trigonometric; solving sets of equations.	X			
2.	Elementary statistics. Uncertainty in numbers representing experimental data, average, standard deviation.	X			X
3.	Units, conversion of units, scaling in powers of 10.	X			X
4.	**Calculus.** Differentiation, integration, limits, slopes, curvature, extrema, series, areas, volumes, graphical presentation of functions from $f(x)$ and from $f''(x)$. Multivariable. A standard set of derivatives and integrals related to functions listed in 1 should be memorized. Inverse relationships. Varying the symbols used for variables. Integration by parts. Exact and inexact differentials; Euler reciprocity relationship for exactness of differentials. Careful specification of which variables are held constant in a partial derivative.		X		X
5.	**Creating, using, and interpreting graphs.**	X	X	X	X
6.	Estimating and making appropriate assumptions. Checking answers for reasonableness, relative importance of terms in equations, appropriate use of linear approximations and Taylor series.	X	X	X	X
7.	Iteration. The Newton methods, gradients, iteration consisting of initialization, successive approximation, and termination steps.		X	X	X
8.	Coordinate systems (Cartesian and polar) and transformations between them. Different frames of reference for coordinate systems.		X		X
9.	Numerical methods. For integration, differentiation, differential equations, finding roots.		X		X
10.	Representation of information. Digital-to-analog and analog-to-digital conversions. Consequences of using different number bases or fractional expressions. Binary, octal, hexadecimal. Enhancement in signal/noise ratio from multiple scanning in which the signal increases linearly with the number of scans, whereas the noise, which is statistical, increases by the square root of the number of scans.		X		X
11.	Statistics. Probability, combinatorics, distributions, uncertainty, confidence intervals, propagation of error.		X		X
12.	Curve fitting. Least-squares methods, regression, using different weights for data, deconvolution (separating out the contributions of several curves of assumed functionality from their overlap in a complicated curve).		X		X
13.	Operators. How they combine and interact. Precedence.		X		X
14.	**Spatial representations.** 3-D geometry, surfaces, projections, slices, perspective.	X	X	X	X

Who Teaches Material*

		1	2	3G	3A
15.	Group theory.				X
16.	**Linear algebra.** Matrix algebra, eigen analysis, basis functions, orthogonality, Fourier series.		X		X
17.	Differential equations. First-order DEs. First-order separable. Partial DEs.		X		X
18.	Symmetry. Transformations. Operators. Symmetric and antisymmetric functions.		X		X
19.	Chemometrics. In its broadest sense chemometrics is the use of mathematical techniques and computational methods to assist the chemical scientist in making and interpreting chemical measurements. Chemometrics includes univariate statistics, multivariate statistics, multivariate modeling (e.g., least-squares regression, partial-least-squares regression), convolution and correlation, pattern recognition, Fourier methods, the calculus, optimization methods (e.g., simplex), and experimental design. The emphasis in recent years has been on extracting useful qualitative and quantitative information from large sets of multivariate data and in developing mathematical models that can predict chemical properties from chemical structure.				X

***1** Material learned in secondary school, which chemists reinforce.

2 Material chemists expect mathematicians to develop.

3G Material in mathematics for which chemists have responsibility at the general chemistry level.

3A Material in mathematics for which chemists have responsibility at the advanced course level.

Concepts with the highest priority are indicated by **boldface.**

APPENDIX B. Understanding and Content, Skills Classification Scheme

Who Teaches Material*

		1	2	3G	3A
1.	Spreedsheets.		X	X	
2.	Graphing.	X	X	X	
3.	Computer algebra (symbolic mathematics by computer).		X		X
4.	Numerical algorithms, iteration.		X		X
5.	Modeling. Analytical (interacting functions such as in a multistep rate process) and molecular (exploring molecular structure and wave functions).			X	X
6.	Graphics software and other visualizations.				X
7.	Algorithms. Understand and apply them (e.g., in Excel and Matlab).		X		X
8.	Reading mathematical expressions and writing them with understanding.	X	X	X	

***1** Material learned in secondary school, which chemists reinforce.

2 Material chemists expect mathematicians to develop.

3G Material in mathematics for which chemists have responsibility at the general chemistry level.

3A Material in mathematics for which chemists have responsibility at the advanced course level.

APPENDIX B. Understanding and Content Skills Classification Scheme

Computer Science

CRAFTY Curriculum Foundations Project
Bowdoin College, October 28–31, 1999

Charles F. Kelemen, Report Editor
Allen Tucker and William Barker, Workshop Organizers

Summary

The general conclusion of the workshop participants is that undergraduate computer science majors need to acquire mathematical maturity and skills, especially in discrete mathematics, early in their college education. The following topics are likely to be used in the first three courses for computer science majors: logical reasoning, functions, relations, sets, mathematical induction, combinatorics, finite probability, asymptotic notation, recurrence/difference equations, graphs, trees, and number systems. Ultimately, calculus, linear algebra, and statistics topics are also needed, but none earlier than discrete mathematics. Thus, such a discrete mathematics course should be offered in the first semester and the prerequisite expectations and conceptual level should be the same as for the Calculus I course offered to mathematics and science majors. Our detailed recommendations respond directly to the series of questions of direct relevance to the CUPM Initiative posed by the Workshop hosts.

While the authors of this report have all been involved in computer science curriculum design in the past, this report does not represent the position of any official ACM or IEEE sanctioned curriculum committee.

At the time of the Bowdoin meeting reported here, a joint task force of the Association for Computing Machinery (ACM) and Computer Society of the Institute for Electrical and Electronic Engineers (IEEE-CS) was beginning to formulate a new set of curriculum recommendations for undergraduate programs in computing. The final report (CC2001) of that joint task force (dated December 15, 2001) is available at:

`www.computer.org/education/cc2001/final/index.htm`

The mathematics recommendations to be found in CC2001 are consistent with those included here from the Bowdoin meeting. A few quotations concerning mathematics from CC2001 are included in Appendix 2.

Narrative

Introduction and Background

We are not aware of any concise, agreed upon, definition of Computer Science. A concise statement that appeals to the editor of this report was given by Jim Foley, Chair of the Computing Research Association Board. "Computer Science discovers and uses the laws of "how to" compute and "how to" organize information to create computational and information artifacts. Computer Science is also concerned with the organization—that is, the architecture—of the physical artifacts that perform computations and that store and transmit information."

Understanding and Content

What conceptual mathematical principles must students master in the first two years?

Students should be comfortable with abstract thinking, notation and its meaning. They should be able to generalize from examples and create examples of generalizations. In order to estimate the complexity of algorithms, they should have a feeling for functions that represent different rates of growth (e.g., logarithmic, polynomial, exponential). In order to reason effectively about the complexity and correctness of algorithms, they should have some facility with formal proofs, especially induction proofs. The same kind of clear and careful thinking and expression needed for a coherent mathematical argument is needed for the design and effective implementation of a computer program [Ralston 84, Henderson 97].

What mathematical problem solving skills must students master in the first two years?

Students should have the mathematical background necessary to be able to represent 'real-world' problem situations with discrete structures such as arrays, linked lists, trees, finite graphs, other multi-linked structures, and matrices. They should be able to develop and analyze algorithms that operate on these structures (e.g., [Cormen 90]). They should understand what a mathematical model is and be able to relate mathematical models to real problem domains (e.g., [Wolz 94, Woodcock 88]). General problem solving strategies such as divide-and-conquer and backtracking strategies are also essential.

What broad mathematical topics must students master in the first two years?

The first three courses for computer science majors are typically an introduction to computer science (containing a large amount of programming), a course in data structures and algorithms, and a course in computer architecture/organization. Some schools put computer architecture before data structures and some do the opposite. A few schools cover discrete mathematics topics before they do much programming [e.g., Baldwin 92, Henderson 90]. The following topics are likely to be used in the first three courses for computer science majors: logical reasoning (propositions, De Morgan's laws, including negation with quantifiers), functions, relations (equivalence relations and partitions), sets, notation ($f: A \to B$; $A \times B$; $A \cap B$) , mathematical induction (structural, strong and weak), combinatorics, finite probability, asymptotic notation (e.g., $O(n^2)$, $O(2^n)$), recurrence/difference equations, graphs and trees, and number systems.

Some examples:

Propositional logic and number systems

A student may have the following code in a program:

```
if ( (i > n) && (a[i] != x) ) do thing1
      else do thing2
```

After some analysis, it is discovered that `thing1` is not necessary at all. The student would like to negate the condition of the if statement and do `thing2` if the negated condition is true; an application of De Morgan's law in propositional logic. This kind of change comes up often in the first two computer science courses. Many students have great difficulty negating a compound logical expression such as the one above.

Computer architecture is usually taught in the first two years of a computer science major. Decimal, binary, and hexadecimal number systems are used extensively. The use of logical expressions and their circuits to realize adders, multiplexors, decoders, etc. are essential for this course. Fluency with the propositional calculus is thus an important prerequisite here too.

Beyond these two examples, a more extended discussion of the centrality of logic in computer science is provided in [Meyers 90, Gries 96].

Growth of functions

In analyzing nested loops, the sum $\sum_{k=1}^{n} k$ occurs often. That this sum evaluates to $n(n+1)/2$ and that as n gets large this sum is quite different from n is important. The sum $\sum_{k=1}^{n} 1/k$ also appears often. That this is approximately $\ln n$ is important. In fact, the notion that $O(\ln n) = O(\log_2 n)$ is also important.

Use of recurrence, induction, and finite probability

One of the best sorting algorithms is called quicksort. Given an array to be sorted, say a[first..last], one array element is chosen as a "pivot". The array is then rearranged by the subprogram partition so that the pivot element is in its proper sorted location and all the elements with value less than or equal to the pivot value are moved to array locations with indices less than the pivot values new index (partdiv). All elements greater than the pivot value are moved to array locations with indices greater than partdiv. Thus the pivot element, now in a[partdiv], is in the correct position for the sorted array. Quicksort then "divides-and-conquers" by making recursive calls to itself to sort both the subarray of elements smaller than a[partdiv] and the subarray of elements larger than a[partdiv]. A version of quicksort is shown below that sorts the integer array a[first..last], with pre- and post-conditions as noted.

```
//Pre-condition: 0 <= first <= last < MAXARRSIZE
//Post-condition:      a[first..last] is in ascending order

void quicksort(IntArr a, int first, int last)
{ int pivotind; //pivot index before partitioning
  int partdiv;   //partition division point after partitioning
  if ((last - first) > 0)  //there is something here to sort
  { pivotind = (first+last)/2;
     //pivot on middle element
    partdiv = partition(a,first, last, pivotind);
     //The partition function returns the index for
     //the sorted position for the pivot element and
     //rearranges the array elements so that upon return
     // a[first..partdiv-1] <= a[partdiv]
     //          < a[partdiv+1..last]
    quicksort(a, first, partdiv-1);
     //sort left part
    quicksort(a, partdiv+1, last);
     //sort right part
  }  // end if
}      // end quicksort
```

In a separate argument, using loop invariants, one can prove that the function partition is correct. **Strong induction** is used to prove quicksort correct. Attempting to prove an incorrectly formulated algorithm correct is often the best way to find out what is wrong with it.

It can be shown that partition takes less than n 'element comparisons' to partition an array of n elements. Using this and assuming that partition always divides the array into equal portions, we get the **recurrence** $T(n) < 2T(n/2) + n$ where $T(n)$ represents the number of 'element comparisons' to quicksort an array of n

elements. If the initial ordering of the array is such that partition divides the array into parts containing 0 and n - 1 elements, then the recurrence for quicksort is $T(n) < T(n-1)+n$. The first case yields $O(n \log n)$ complexity, while the second yields $O(n^2)$. Being able to derive recurrences of this sort and to solve them is important in early computer science courses. Students should also be able to analyze the expected performance of quicksort. If all orderings of the initial array are equally likely, the **expected performance** is $O(n \log n)$ and the constant hidden in the big-oh is small enough that quicksort is preferable to many other sorting algorithms whose worst-case performance is $O(n \log n)$. Thus, **probabilistic analysis** is important.

Binary search trees are important data structures covered in a second computer science course. They are most easily defined using **recursive definitions** and most easily processed using recursive algorithms. For example, an inorder traversal of a binary search tree is easily expressed recursively but extremely difficult to code without using recursion. Many algorithms on binary search trees depend upon the height of the tree. Results relating the height of the tree to the number of nodes in the tree are most easily proved using **induction**.

Mathematics in the rest of the computer science curriculum

Many intermediate and advanced computer science courses use mathematical topics that we would like students to master.

- Scientific computing and numerical analysis use differential and integral calculus, multidimensional calculus, and linear algebra.
- Computer graphics uses linear algebra (matrix algebra, change of coordinates), 3-dimensional calculus, and topics from geometry. Theory of Computation and Algorithms courses use induction and diagonalization proofs. Counterexamples and proof by contradiction are important.

More advanced mathematical topics may also be used in select upper-division computer science courses.

- Transforms are used in speech understanding and synthesis algorithms.
- Wavelets are used in compression algorithms.
- Number theory, group and ring theory are used in encryption algorithms.

The Computing Sciences Accreditation Board [CSAB 99] recommends the following for undergraduate computer science majors: "The curriculum must include at least one-half year [4 or 5 courses] of mathematics. This material must include discrete mathematics, differential and integral calculus, and probability and statistics, and may include additional areas such as linear algebra, numerical analysis, combinatorics, and differential equations." Similar recommendations appear in the ACM/IEEE Curriculum 91 Report [ACM/IEEE 91] and the Liberal Arts Model Curriculum [Walker 96, Gibbs 86], which are widely-used models for designing undergraduate computer science major programs in the United States.The GRE in computer science [GRE 99] weights 25% on theory and 15% on mathematical background. Theory depends heavily on discrete mathematics.

Topics listed under mathematical background include:

- Discrete Structures
  1. Mathematical logic
  2. Elementary combinatorics, including graph theory and counting arguments
  3. Elementary discrete math with number theory, discrete probability, recurrence relations
- Numerical mathematics
  1. Computer arithmetic with number representations, roundoff errors, overflow, underflow
  2. Classical numerical algorithms
  3. Linear algebra

What priorities exist among these topics?

For the early computer science courses, discrete mathematics topics take priority over calculus and linear algebra [Ralston 84]. If these discrete mathematics topics are not covered in a first- or second-semester mathematics course they must be introduced in the computer science courses themselves. This slows down the computer science course and probably leads to a more cursory treatment of the mathematics topics than might be appropriate in a mathematics course. Given the current difficulty in hiring computer science faculty, we suspect that most computer science departments would welcome a freshman-level discrete mathematics course covering the topics needed for computer science, but taught by the mathematics department. In fact, many computer science departments consider these topics so important that they offer their own courses covering them. Some of these courses bear titles like "Discrete Structures" or "Computational Structures." (E.g., see [Epp 95, Gersting 99, Rosen 99] for a sampling of contemporary discrete mathematics texts.)

What is the desired balance between theoretical understanding and computational skill?

We think both theoretical understanding and computational skill are important. Computational skill (in the sense of plug and chug) is less important than the ability to recognize when these topics may be used productively in algorithmic problem solving and computational modeling. On the other hand, we would really like students to be able to formulate and complete induction proofs. If this is considered computational, then computational skill is very important.

What are the mathematical needs of different student populations and how can they be fulfilled?

Computer science courses for humanities students do not require sophisticated mathematics. Computer science courses specifically designed for business majors are well served by the business mathematics courses. Some colleges and universities offer special computer science courses for science and engineering majors. These students have such heavy mathematics and science requirements in the first two years that it is probably not possible to require them to take a discrete mathematics course early. Covering some discrete mathematics topics (say induction and propositional logic) in Calculus I would be helpful for these computer science courses. Ideally, for computer science majors, discrete mathematics should be covered before Calculus.

Often the first two computer science courses for computer science majors are also taken by majors from mathematics, the natural sciences, economics, social sciences, and others who want to gain a deeper mastery of this important field. For many of these students, a first semester discrete mathematics course would be of value.

Technology

How does technology affect what mathematics should be learned in the first two years?

We support the goal of FITness (Fluency in Information Technology) promulgated in the NAS report, "Being Fluent with Information Technology" [NAS 99]. The key idea is that students of all disciplines should learn enough foundational material in their formal education that they can embark on "a process of lifelong learning in which individuals continually apply what they know to adapt to change and acquire more knowledge to be more effective at applying information technology to their work and personal lives." In other words, everyone needs more than a superficial acquaintance with technology as a tool in their own areas of interest. Computer technology should be incorporated deeply and thoroughly into all mathematics curricula.

Instructional Techniques

What instructional methods best develop the mathematical comprehension needed for your discipline?

We have no easy answers here. The following methods seem to work well in computer science and we would presume that they might work well in mathematics: interactive collaborative learning leading to team/group reports, peer learning/teaching, learning center/laboratories (staffed), encouraging high-quality public written and oral communications, and providing research opportunities.

Instructional Interconnections

What impact does mathematics education reform have on instruction in your discipline?

We feel that the migration of Calculus toward problem solving (from "plug and chug") is good, though less relevant in impact on computer science than similar reforms in Discrete Math might be. The mathematics community's inattention to Discrete Math early has forced many computer science departments to assimilate and teach these topics themselves.

How should education reform in your discipline affect mathematics instruction?

The use of labs, group work, and peer learning has proven very beneficial in computer science education [Parker 90]. We suspect that the use of these techniques would be productive in some mathematics courses, especially discrete mathematics courses (e.g. [Epp 95]).

How can dialogue on educational issues between your discipline and mathematics best be maintained?

A joint IEEE Computer Society/ACM Task Force on the "Year 2001 Model Curricula for Computing" [ACM/IEEE 01] has been formed "to review the 1991 curricula and develop a revised and enhanced version for the Year 2001 that addresses developments in computing technologies in the past decade and will sustain through the next decade." We hope that the CUPM committee will be able to interact with this computer science curriculum planning group. Other forums might be MAA and SIGCSE conferences and articles and newsletters of these organizations.

REFERENCES

[ACM/IEEE 91] Tucker, A., et al. "A Summary of the ACM/IEEE-CS Joint Task Force Report 'Computing Curricula 1991'". Communications of the ACM, June 1991. pp. 69–84. For the full report, including the mathematics requirements for computer science majors, see www.acm.org/education/curr91/homepage.html.

[ACM/IEEE 01] "Year 2001 Model Curricula for Computing" computer.org/education/cc2001/index.htm.

[Baldwin 92] D. Baldwin and J. A. G. M. Koomen. "Using Scientific Experiments in Early Computer Science Laboratories". Proceedings of the Twenty-Third SIGCSE Technical Symposium on Computer Science Education, March 1992 (SIGCSE Bulletin, Mar. 1992). pp. 102–106.

[Cormen 90] Thomas H. Cormen, Charles E. Leiserson, and Ronald L. Rivest. Introduction to Algorithms, MIT Press, Cambridge, Massachusetts, 1990, pp. 64–72.

[CSAB 99] Computing Sciences Accreditation Board, Revised CSAC Evaluative Criteria, www.csab.org.

[Epp 95] Susanna Epp. Discrete Mathematics with Applications, 2nd Edition, PWS Publishing, Boston, Massachusetts, 1995.

[Gersting 99] Judith L. Gersting. Mathematical Structures for Computer Science, Fourth Edition, W.H. Freeman, 1999.

[Gibbs 86] Gibbs, N. and A. Tucker. "A Model Curriculum for a Liberal Arts Degree in Computer Science", Communications of the ACM, Mar. 1986. pp. 202–210.

[GRE 99] Graduate Record Exams free publications, ftp://etsis1.ets.org/pub/gre/275516.pdf.

[Gries 96] Gries, D. "The Need for Education in Useful Formal Logic", IEEE Computer, April 1996, pp. 29–30.

[Henderson 90] Henderson, P.B., "Discrete Mathematics as a Precursor to Computer Programming," Proceedings of the Twentieth SIGCSE Technical Symposium on Computer Science Education, Washington, DC, February 1990.

[Henderson 97] Henderson, P.B., "Problem Solving, Discrete Mathematics and Computer Science," DIMACS Series on Discrete Mathematics and Theoretical Computer Science, Vol. 36, American Mathematical Society, 1997.

[Meyers 90]. J.P. Meyers. "The Central Role of Mathematical Logic in Computer Science", ACM SIGCSE Bulletin. 22(1), pp 22–26, 1990.

[NAS 99] National Academy of Sciences, "Being Fluent with Information Technology", National Academy Press (1999), books.nap.edu/catalog/6482.html.

[Parker 90]. J. Parker, R. Cupper, C. Kelemen, R. Molnar, G. Scragg. "Laboratories in the Computer Science Curriculum", Computer Science Education 1, pp. 205–221, 1990.

[Ralston 84]. Anthony Ralston. "The First Course in Computer Science Needs a Mathematics Corequisite", Communications of the ACM, 27(10), pp 1002–1005, 1984.

[Rosen 99] Kenneth Rosen. Discrete Mathematics and its Applications, Fourth Edition, McGraw-Hill, 1999

[Stavely 99] Allan M. Stavely. "High-Quality Software Through Semiformal Specification and Verification." Proc. 12th Conference on Software Engineering Education and Training (CSEE&T'99), March 22–24, 1999, New Orleans.

[Walker 96]. Walker, H. and G. M. Schneider. "A Revised Model Curriculum for a Liberal Arts Degree in Computer Science", Communications of the ACM, Dec. 1996. pp. 85–95.

[Wolz 94]. Wolz, U. and E. Conjura. "Integrating Mathematics and Programming into a Three Tiered Model for Computer Science", ACM SIGCSE Bulletin. 26(1), pp 223–227, 1994

[Woodcock 88] Woodcock, J. and M. Loomes. Software Engineering Mathematics, Addison-Wesley, 1988.

WORKSHOP PARTICIPANTS

Computer Science Participants
(See the Appendix for detailed biographies.)

Owen Astrachan, Associate Professor of the Practice and Director of Undergraduate Studies for Teaching and Learning in Computer Science, Duke University.

Doug Baldwin, Associate Professor of Computer Science, State University of New York at Geneseo.

Kim B. Bruce, Frederick Lattimer Wells Professor of Computer Science, Williams College.

Peter B. Henderson, Professor and Head, Department of Computer Science and Software Engineering, Butler University.

Charles F. Kelemen, Professor of Computer Science, Swarthmore College.

Dale Skrien, Professor of Computer Science, Colby College.

Allen B. Tucker, Anne T. and Robert M. Bass Professor of Natural Sciences in the Department of Computer Science, Bowdoin College.

Charles F. Van Loan, Chair of the Department of Computer Science, Cornell University.

Mathematics Participants

Thomas Banchoff, Professor of Mathematics, Brown University.

William Barker, Professor of Mathematics, Bowdoin College.

Thomas Berger, Professor of Mathematics, Colby College.

Susan Ganter, Associate Professor of Mathematical Sciences, Clemson University.

Deborah Hughes Hallett, Professor of Mathematics, University of Arizona.

Harvey Keynes, Professor of Mathematics, University of Minnesota.

William McCallum, Professor of Mathematics, University of Arizona.

Donald Small, Professor of Mathematics, U.S. Military Academy, West Point.

ACKNOWLEDGEMENTS

We would like to thank CUPM, CRAFTY, and the local organizers of the Bowdoin workshop (William Barker, Guy Emery, Allen Tucker, and Katharine Billings) for a very pleasant and enlightening experience.

APPENDIX 1 : Biographies of the Workshop Participants

Owen Astrachan is Associate Professor of the Practice and Director of Undergraduate Studies for Teaching and Learning in Computer Science at Duke University. He has an AB in mathematics from Dartmouth College (1978) and MAT (1979), MS (1989), and PhD (1992) degrees in computer science from Duke University. He has published in the areas of automated reasoning, parallel programming, and computer science education. He has served as Chief Reader for Advanced Placement Computer Science and chaired College Board/SIGCSE committees making recommendations to the AP program about language changes. He is the PI or co-PI on more than one million dollars of NSF-funded sponsored research including a CAREER award bridging software engineering and computer science education. He has won teaching awards at both Duke and the University of British Columbia.

Doug Baldwin is associate professor of computer science at the State University of New York at Geneseo, where he has led the development of a mathematically rigorous introductory computer science course sequence. He is the author of a number of papers, and recipient of a number of grants, in computer science education. He holds BS, MS, and Ph. D. degrees in computer science from Yale University. He is a member of the ACM, and of the IEEE Computer Society, and served on the 1998 and 1999 program committees for the annual symposium of the ACM's Special Interest Group on Computer Science Education.

Kim B. Bruce is Frederick Lattimer Wells Professor of Computer Science at Williams College, and has served as department chair several times. He received his BA at Pomona College and MA and Ph.D. in Mathematical Logic at the University of Wisconsin. He has been a visiting professor or visiting scientist at Princeton University, Stanford University, MIT, the Ecole Normale Superieure in Paris, the University of Pisa, and the Newton Institute for Mathematical Sciences at Cambridge University. He has published widely in the area of the semantics and design of programming languages as well as computer science education. He was a contributor to the ACM/IEEE-CS Joint Curriculum Task Force that developed Computing Curricula 1991, and participated in the design of the original 1986 Liberal Arts Model Curriculum in Computer Science and its revision in 1996. He currently chairs the advisory committee on programming languages for the ACM/IEEE computer science Curriculum 2001 effort, and is workshop chair for the series of workshops on Foundations of Object-Oriented Languages. He is a "Golden Core" member of IEEE-CS and has received ACM Meritorious Service and Recognition of Service awards.

Peter B. Henderson is professor and head of the Department of Computer Science and Software Engineering at Butler University in Indianapolis. For the previous 25 years he was at SUNY Stony Brook working in the areas of software engineering, programming environments and computer science education. He received his B.S. and M.S. degrees in Electrical Engineering from Clarkson College and his Ph.D. from Princeton University. Professor Henderson has chaired three SIGSOFT/SIGPLAN Symposiums on Software Development Environments and has numerous publications in computer science and mathematics education.

Charles F. Kelemen is Professor and Chair of Computer Science at Swarthmore College where he holds the Edward Hicks Magill Professorship of Mathematics and Natural Sciences. He earned a BA in Mathematics from Valparaiso University and an MA and PhD from the Pennsylvania State University. Before coming to Swarthmore College, he held regular faculty positions at Ithaca College and LeMoyne College and visiting positions in Computer Science at Cornell University. In 1985, Kelemen founded the Computer Science Program at Swarthmore College and chaired it from 1985 until 1999. He participated in the design of the original 1986 Liberal Arts Model Curriculum in Computer Science and its revision in 1996 and was a reviewer of the ACM/IEEE-CS Joint Computing Curricula 1991. He has published books, research, and educational articles in both mathematics and computer science. He is a member of ACM, IEEE-CS, CPSR, MAA, and the Liberal Arts Computer Science Consortium (LACS).

Dale Skrien is a Professor of Computer Science at Colby College. He has a BA in mathematics from St. Olaf College, an MS and PhD in mathematics from the University of Washington (1980), and an MS in computer science from the University of Illinois (1985). He has taught mathematics and computer science at Colby College since 1980. He has also been a Fulbright lecturer at the University of Malawi and a software engineer contractor for Digital Equipment Corporation. His research interests include graph theory, computer music, and computer science education.

Allen B. Tucker is the Anne T. and Robert M. Bass Professor of Natural Sciences in the Department of Computer Science at Bowdoin College. He has held similar positions at Geogetown and Colgate Universities. He has a BA in mathematics from Wesleyan University (1963) and an MS and PhD in computer science from Northwestern University (1970). He has various publications in the areas of programming languages, natural language processing, and computer science education. Professor Tucker co-chaired the ACM/IEEE-CS Joint Curriculum Task Force that developed Computing Curricula 1991 and is co-author of the 1986 Liberal Arts Model Curriculum in Computer Science. He is a Fellow of the ACM, and has been a member of the IEEE Computer Society, Computer Professionals for Social Responsibility, and the Liberal Arts Computer Science (LACS) Consortium.

Charles F. Van Loan is Chair of the Department of Computer Science at Cornell University where he is the Joseph C. Ford Professor of Engineering. He received his BS (1969), MS (1970), and PhD (1973) in Mathematics from the University of Michigan. He works in the area of matrix computations and has written several textbooks including Matrix Computations (with G.H. Golub), Computational Frameworks for the Fast Fourier Transform, Introduction to Scientific Computing, and Introduction to Computational Science and Mathematics. His papers “Computer Science and the Liberal Arts Student” and “Building Freshman Intuition for Computational Science” reflect his interest in undergraduate education. See `www.cs.cornell.edu/cv/`.

APPENDIX 2: Some Quotes from Curriculum 2001

This material is being added as this report goes to press in October, 2002. It was not available at the time of the Bowdoin meeting. The full ACM, IEEE-CS report is available at:

`www.computer.org/education/cc2001/final/index.htm`

From Section 7.4 Integrating discrete mathematics into the introductory curriculum.

As we discuss in , the CC2001 Task Force believes it is important for computer science students to study discrete mathematics early in their academic program, preferably in the first year. There are at least two workable strategies for accomplishing this goal:

From Appendix A: The CS body of knowledge.

DS. Discrete Structures (43 core hours)

DS1. Functions, relations and sets (6)
DS2. Basic Logic (10)
DS3. Proof Techniques (12)
DS4. Basics of Counting (5)
DS5. Graphs and trees (4)
DS6. Discrete Probability (6)

From Appendix B: CS Course Descriptions

CS115. *Discrete Structures for Computer Science*

Offers an intensive introduction to discrete mathematics as it is used in computer science. Topics include functions, relations, sets, propositional and predicate logic, simple circuit logic, proof techniques, elementary combinatorics, and discrete probability.

Prerequisites: Mathematical preparation sufficient to take calculus at the college level.

Syllabus:

- *Fundamental structures:* Functions (surjections, injections, inverses, composition); relations (reflexivity, symmetry, transitivity, equivalence relations); sets (Venn diagrams, complements, Cartesian products, power sets); pigeonhole principle; cardinality and countability
- *Basic logic:* Propositional logic; logical connectives; truth tables; normal forms (conjunctive and disjunctive); validity; predicate logic; limitations of predicate logic; universal and existential quantification; modus ponens and modus tollens
- *Digital logic:* Logic gates, flip-flops, counters; circuit minimization
- *Proof techniques:* Notions of implication, converse, inverse, contrapositive, negation, and contradiction; the structure of formal proofs; direct proofs; proof by counterexample; proof by contraposition; proof by contradiction; mathematical induction; strong induction; recursive mathematical definitions; well orderings
- *Basics of counting:* Counting arguments; pigeonhole principle; permutations and combinations; recurrence relations
- *Discrete probability:* Finite probability spaces; conditional probability, independence, Bayes' rule; random events; random integer variables; mathematical expectation.

Notes: This implementation of the Discrete Structures area (DS) compresses the core material into a single course. Although such a strategy is workable, many institutions will prefer to use two courses to cover this material in greater depth. For an implementation that uses the two-course model, see the descriptions of CS105 and CS106.

CS105. *Discrete Structures I*

Introduces the foundations of discrete mathematics as they apply to computer science, focusing on providing a solid theoretical foundation for further work. Topics include functions, relations, sets, simple proof techniques, Boolean algebra, propositional logic, digital logic, elementary number theory, and the fundamentals of counting.

Prerequisites: Mathematical preparation sufficient to take calculus at the college level.

Syllabus:

- *Introduction to logic and proofs:* Direct proofs; proof by contradiction; mathematical induction
- *Fundamental structures:* Functions (surjections, injections, inverses, composition); relations (reflexivity, symmetry, transitivity, equivalence relations); sets (Venn diagrams, complements, Cartesian products, power sets); pigeonhole principle; cardinality and countability
- *Boolean algebra:* Boolean values; standard operations on Boolean values; de Morgan's laws
- *Propositional logic:* Logical connectives; truth tables; normal forms (conjunctive and disjunctive); validity
- *Digital logic:* Logic gates, flip-flops, counters; circuit minimization
- *Elementary number theory:* Factorability; properties of primes; greatest common divisors and least common multiples; Euclid's algorithm; modular arithmetic; the Chinese Remainder Theorem
- *Basics of counting:* Counting arguments; pigeonhole principle; permutations and combinations; binomial coefficients

CS106. *Discrete Structures II*

Continues the discussion of discrete mathematics introduced in CS105. Topics in the second course include predicate logic, recurrence relations, graphs, trees, matrices, computational complexity, elementary computability, and discrete probability.

Prerequisites: CS105

Syllabus:

- *Review* of previous course
- *Predicate logic:* Universal and existential quantification; modus ponens and modus tollens; limitations of predicate logic
- *Recurrence relations:* Basic formulae; elementary solution techniques
- *Graphs and trees:* Fundamental definitions; simple algorithms ; traversal strategies; proof techniques; spanning trees; applications
- *Matrices:* Basic properties; applications
- *Computational complexity:* Order analysis; standard complexity classes
- *Elementary computability:* Countability and uncountability; diagonalization proof to show uncountability of the reals; definition of the P and NP classes; simple demonstration of the halting problem
- *Discrete probability:* Finite probability spaces; conditional probability, independence, Bayes' rule; random events; random integer variables; mathematical expectation

Notes: This implementation of the Discrete Structures area (DS) divides the material into two courses: CS105 and CS106. For programs that wish to accelerate the presentation of this material, there is also CS115, which covers the core topics in a single course. The two-course sequence, however, covers some additional material that is not in the compressed version, primarily in the Algorithms and Complexity area (AL). As a result, the introductory course in algorithmic analysis (CS210) can devote more time to advanced topics if an institution adopts the two-course implementation.

Like CS105, this course introduces mathematical topics in the context of applications that require those concepts as tools. For this course, likely applications include transportation network problems (such as the traveling salesperson problem) and resource allocation.

Engineering: Chemical Engineering

CRAFTY Curriculum Foundations Project
Clemson University, May 4–7, 2000

Michael D. Graham, Report Editor
Kenneth Roby and Susan Ganter, Workshop Organizers

Summary

It is clear that application of mathematical concepts and the generation of mathematical solutions to engineering problems are essential to the educational programs of all undergraduate engineering students. Enhanced cooperation between the mathematics faculty and the engineering faculty can lead to a better experience for our students.

Without exception, this group felt that the workshop was a very productive way to promote dialogue between the mathematics and engineering education communities and we would like to see workshops of this type continue to be held. Another venue that mathematics educators may want to explore is the American Society for Engineering Education (www.asee.org), which has a mathematics division. On the other hand, it may be productive for engineering educators to attend MAA meetings. Perhaps most importantly, mechanisms need to be implemented to promote interaction between engineering and mathematics faculty within individual universities — good relationships at this level will enable mathematics faculty to understand what material the engineering faculty would like to see reinforced and emphasized, as well as enabling engineering faculty to gain a better understanding of the issues surrounding mathematical preparation of entering freshman engineering majors.

Narrative

Introduction and Background

What are Chemical Engineers? Since this is a report for mathematicians, we thought an appropriate introduction would be to try to say what chemical engineers do, why we need mathematics, and how we use it. A reasonably broad definition of what we do is that chemical engineers design materials and the processes by which materials are made. Traditionally, chemical engineers have been associated with the petroleum and large-scale chemical industries, but especially in recent years, chemical engineers have been involved in pharmaceuticals, foods, polymers and materials, microelectronics and biotechnology. The core subjects that underlie and unify this broad field are thermodynamics, chemical reaction processes, transport processes (i.e., the spatial and temporal distribution of mass, momentum and energy) and process dynamics, design and control.

On top of this fundamental framework, a central emphasis of chemical engineering education is model building and analysis. A good chemical engineer brings together the fundamentals to build and refine a mathematical model of a process that will help him or her understand and optimize its performance. To be good at model building and analysis, students must have at hand the mathematical background to understand and work with the core scientific areas, as well as to find solutions to the final model that they build. In this context, the "solution" to a mathematics problem is often in the understanding of the behavior of the process described by the mathematics, rather than the specific closed form (or numerical) result.

Here's an example. A starting point for understanding any process is writing down the conservation laws that the system or process satisfies: for conserved quantities, accumulation = input – output. Depending on the level of detail of the model, this equation might be, for example, a large set of linear algebraic equations that determine the relationships between fluxes of chemical species throughout the process (a species balance), or it might be a set of parabolic partial differential equations governing the temperature and composition of the fluid in a chemical reactor. In the thermodynamics of multiphase systems, energy is conserved but takes on a variety of forms; a good knowledge of multivariable differential calculus is essential here to keep track of everything.

Understanding and Content

Mathematics for Chemical Engineering. We do not view our role here as one of prescribing the mathematics curriculum—we do not want mathematics instruction to provide only what students can "get by" with knowing. Nor do we want to come down on either side of the "traditional" vs. "reform" debate—it is likely that both sides are right, to an extent. We have collected here some general thoughts on subject matter and emphasis that arose in our discussions:

Precalculus Foundations. By foundations, we mean

- basic knowledge of families of functions (polynomial, exponential...) in terms of data, graphs, words and equations, basic trig identities, properties of logarithms,
- equations, inequalities,
- basic logic and algorithms,
- small linear systems of equations,
- coordinate systems,
- basic arithmetic and manipulation skills.

Mastery of the above areas is crucial. Probably the most important thing the mathematics education community can do here is to actively critique the pedagogy of K–12 education – to help sort out which "reforms" are productive from those that are merely ed-school fads and to encourage schools not to neglect the education of the more mathematically inclined students by focusing the curriculum too narrowly on the average performer. Another important role here is to provide programs that help K–12 mathematics teachers understand something about the applications of the mathematics that they teach (engineering faculty should do much more here).

Linear Mathematics. We feel that our students would benefit from earlier exposure to the basics of linear systems in R^n, particularly

- the geometry of linear spaces,
- vector algebra (especially in 3D),
- $\mathbf{Ax} = \mathbf{b}$ (existence and uniqueness, Gaussian elimination, geometric interpretation, over- and under-determined systems and least squares problems),
- $\mathbf{Ax} = \mathrm{l}\mathbf{x}$ (characteristic polynomial and diagonalization, Jordan form, range and nullspace of $\mathbf{A}$, geometry).

At Wisconsin, there is a course on "linear mathematics", which introduces these notions and applies them to systems of ordinary differential equations (see below). Many chemical engineering students take this in lieu of the traditional differential equations class.

Calculus and Differential Equations. In our discussions of calculus, the importance of visualization repeatedly arose, especially as a guide to differential and vector calculus in multiple dimensions, plotting (e.g., what function is linear on a log-log plot?), working in cylindrical and spherical coordinate systems and how to convert between coordinate systems. Though students must learn techniques such as integration by parts, somewhat less time could be spent on techniques for evaluating complicated integrals. The time saved could be spent on topics such as visualizing the application of the chain rule in multiple dimensions. Understanding of truncated Taylor series for local approximation of functions is very important and should be seen early and often. In differential equations, a thorough knowledge of linear constant coefficient systems (IVPs and BVPs; see above) is preferable to emphasis on existence theory and series solutions for non-constant coefficient problems. Some qualitative theory for nonlinear systems would be nice.

Probability and Statistics. Alumni surveys typically show that probability and statistics, in addition to the extensive use of spread-sheeting software, is the most common application of mathematics for the practicing chemical engineer with a Bachelor of Science degree. Key issues here include parameter estimation, experimental design, sampling and the origins and properties of various distribution functions.

Students interested in graduate school should be encouraged by their mathematics professors as well as their engineering advisors to take additional mathematics courses. A final general comment: for engineers, concepts are more important than proofs, but students should have some idea of the power of a theorem. In other words, we are comfortable with students learning mathematical facts without necessarily having seen the proof.

Technology and Instructional Techniques

A fair amount of the discussion at the workshop, within our group and others, centered around the use of "technology" in the mathematics courses for engineers. In the discussions, "technology" meant a number of different things, from numerical methods to graphing calculators to symbolic manipulation packages. We'd like to emphasize here some points to be kept in mind when thinking of the introduction of these tools into mathematics courses. So here are two simple but common questions and our responses:

"Why should I learn to do it by hand?"

- sense of form of mathematical expressions, understanding of what manipulations are available, facility with these manipulations
- fluency in the language of mathematical concepts,
- appreciation and recognition of mathematical rigor,
- discipline, maturity, confidence of mastery,
- closed form results are best, if available,
- recognition of limitations of closed form results, where things get difficult,
- knowledge of what computers do

"Use of computers dumbs down the mathematics course—why use them?"

- solution of realistic (complex) problems, many of which involve numerical solutions—In upper level courses, extensive use is made of programs like *MATLAB*™, *MathCad*™, *Mathematica*™, *Polymath*™, and *Octave.* (GNU *Octave* is freely available at `www.che.wisc.edu/octave`.)
- efficient exploration of solution and design space
- visualization, especially in multidimensional and vector calculus
- relief from tedium
- confidence in results derived by hand

Ultimately, we feel that the technology should take a back seat in mathematics courses until it becomes necessary to solve interesting problems. For example, in a linear algebra course, students should be able to do LU decomposition of a 3×3 system by hand before they are shown that *MATLAB* does it in one command. At the same time, it is useful to point out the relationship between numerical techniques and exact ones (e.g., a Riemann sum can be evaluated numerically to approximate an integral). Students should have a solid understanding regarding limitations of numerical methods and their accuracy. They should clearly see the power of analytical solutions when such solutions can be found.

Instructional Interconnections

A suggestion for coupling mathematics and engineering education. One set of issues that arose repeatedly in the workshop discussions was the concern that students don't see connections between mathematical tools, concepts and principles and their wide utility in engineering. A related concern was the time lag between exposure to mathematics and its application and importance in the solution of real engineering problems. The notion of "just-in-time" learning arose repeatedly, and the suggestion was made that the mathematics courses be more application- or example-driven and be more evenly spread through the curriculum, rather than "front loaded" into the first two years. Our group shares these concerns, but also feels that:

(1) part of the beauty and power of mathematics is that it is example-independent—calculus applies to economics just as it does to mechanics,

(2) the time spent developing the background for engineering applications is time not spent on mathematical principles and tools, and

(3) a straightforward "just-in-time" approach will not satisfy all the engineering majors—electrical engineers do not need Laplace transforms at the same time as chemical engineers.

We propose that an alternate structure be considered for addressing these concerns, which are essentially about how to connect mathematics and engineering in the students' minds. We suggest the introduction of **discipline specific supplements**, especially for the calculus sequence. These could be workbooks or web pages, for example containing

- engineering background material, e.g., some basic thermodynamics, and how specific mathematical principles and/or tools (e.g., total differentials, partial derivatives in several dimensions) can be applied,
- exercises or projects integrating mathematics and engineering,
- additional discipline-specific emphasis, e.g. trigonometric identities and manipulations for electrical engineering students.

These could be used independently by the students, or used in a one-credit course running in parallel with the calculus courses, or simply be resources for mathematics instructors wishing to gain perspective on engineering applications or bring engineering applications into the mathematics classroom. This is perhaps overly ambitious, but we believe it is worth considering. Within chemical engineering, there is an organization called CACHE (Computer Aids for Chemical Engineering—`www.CACHE.org`) that may take a role in studying this possibility. Michael B. Cutlip of this working group will transmit this opportunity to the CACHE organization for possible coordination with the MAA.

WORKSHOP PARTICIPANTS

Jenna P. Carpenter, Interim Academic Director, Chemical Engineering, Civil Engineering and Geosciences, Louisiana Technological University, Ruston, LA

Michael B. Cutlip, Professor of Chemical Engineering and Director of the Honors Program, University of Connecticut, Storrs CT

Michael D. Graham, Associate Professor of Chemical Engineering, University of Wisconsin-Madison, Madison WI

Anton J. Pintar, Associate Professor of Chemical Engineering, Michigan Technological University, Houghton MI

Jan A. Puszynski, Professor of Chemistry and Chemical Engineering, South Dakota School of Mines and Technology, Rapid City, SD

Mathematics Participants

James Lightbourne, Senior Science Advisor, Division of Undergraduate Education, National Science Foundation

Don Small, Professor, Mathematical Sciences, U.S. Military Academy, West Point

ACKNOWLEDGEMENTS

We are grateful to Professors J. B. Rawlings and W. H. Ray (UW-Madison) and J. F. Brady (Caltech) and to Sangtae Kim (Vice President for Scientific Information Resources, Parke-Davis) for their critical reading and insightful comments on an earlier version of this report.

Engineering: Civil Engineering

CRAFTY Curriculum Foundations Project
Clemson University, May 4–7, 2000

Lynn Katz, Report Editor
Kenneth Roby and Susan Ganter, Workshop Organizers

Summary

This report addresses the mathematical knowledge and skills required for civil engineering students in their introductory courses. One of the key elements identified throughout this document has been the need to integrate mathematics and engineering curricula so that students apply their mathematical skills soon after the material is taught. In order to accomplish this goal, it is necessary to maintain interactions between mathematics and engineering faculty and it may be advisable for introductory mathematics courses to be provided over a three or four year time period rather than a two year period.

In addition, it is essential for mathematics curriculum to focus on developing problem-solving skills. This can be accomplished by adjusting the balance of course material to more applied and numerical solution techniques and less advanced analytical techniques, integrating more technology into the curriculum, and coordinating the mathematics and engineering curricula. Our goal is not to dilute the introductory mathematics curriculum but rather to provide more in-depth understanding and applications of basic concepts prior to introducing more complex topics. The net effect of this approach will be that students gain a greater appreciation of mathematics and will be encouraged to pursue these more complex concepts in advanced courses.

Narrative

Introduction and Background

Civil Engineering is a field of broad scope that ranges from the design and construction of structures, roadways and pollution control processes to the management of our natural and engineered resources. The breadth of the field has led to division into a number of subdisciplines including structural engineering, geotechnical engineering, transportation engineering, environmental engineering, materials, construction management, and water resources engineering. Civil engineering draws on a number of science disciplines including chemistry, physics, ecology, geology, microbiology, material science, economics and mathematics, and statistics.

The level of mathematical and statistical skills required of civil engineering professionals varies with the type of work and the subdiscipline. Many of the technical mathematics skills required for our students are needed primarily to understand the concepts introduced in their engineering courses. The majority of our graduates employ very little of the technical mathematics skills that they learned in college on a day to day basis; however, they rely heavily on their problem solving skills. Very few of our students graduating with Bachelor of Science degrees code new software or solve problems analytically, yet they rely on very sophisticated software and tools that use all levels of advanced mathematics. Still, a significant fraction of our graduates utilize advanced numerical modeling and statistical analyses to evaluate data and predict performance. It is at the graduate level where advanced mathematics is used most heavily. In addition, one of the most significant trends in the past two decades has been the reliance on spreadsheets and commercially available software packages for solving many routine mathematical problems.

This report summarizes the panel's evaluation of the curriculum needs for introductory mathematics courses for civil engineering students. Our recommendations focus on several **key themes** that we believe are essential for improving the quality of mathematics and engineering education:

- A strong emphasis should be given to developing problem-solving skills across the curriculum.
- The introduction of engineering content should start early in the curriculum and the timing of mathematics curriculum content should be closely linked to its use in engineering courses.
- Introductory mathematics content should focus on developing a sound understanding of key fundamental concepts and their relevance to applied problems.
- A stronger emphasis should be placed on numerical solution techniques (e.g., root finding, interpolation, curve fitting, numerical differentiation and integration).
- Curriculum reform requires interdisciplinary coordination between provider and end-user departments.

These themes are re-iterated and expounded upon as we address several key areas of concern for mathematics curriculum. These areas include: understanding and content, technology, instructional interconnections, and instructional techniques.

Understanding and Content

There are a number of early conceptual mathematical principles that are required for civil engineering students. The range of topics is rather broad and covers the major concepts in algebra, trigonometry, logarithms, graphical analysis, data transformation, systems of equations, vectors, series, matrix algebra, integration, differentiation, basic differential equations, probability, statistics and optimization. A complete list of these topics is included in the Appendix.

It may be beneficial to evaluate the order of topics in order to arrange for them to coincide more closely with their use in engineering and basic science courses, especially physics. For example, basic numerical methods and matrix algebra are often stressed very early in the engineering curriculum. In contrast, multivariable calculus is not used until much later. If students focus on gaining a better understanding of basic principles and have a chance to apply them in their engineering classes before they are introduced to more complex topics, they may gain a greater appreciation of mathematics and maintain their enthusiasm for learning new concepts.

The major difference between the needs of current and past civil engineers relates to the techniques available for solving complex problems. Whereas in the past it was necessary to provide in-depth treatment of many different analytical techniques for *solving problems* in each of these areas, the advent of calculators, computers, and user-friendly software packages that replaced the need for broad instruction in numerous solution techniques that are now rarely used in practice. Rather, introductory mathematics instruction should concentrate on teaching the major concepts of each of these areas, their physical meaning and their application to solving realistic problems.

In addition, it is extremely important that students develop the ability to recognize the application of these concepts to applied problems. For example, we believe that it is important for students to understand that integration can be used to determine the area under a curve, and that differentiation can be used to determine the slope of a curve at a particular location. Furthermore, when students are faced with a problem in which they must determine the area under a curve they should be able to recognize that they must integrate and that there are both analytical and numerical tools available to accomplish this task. In contrast, it is less important for students to learn the multitude of analytical techniques for integrating, especially techniques that apply to complex problems that are more likely to be solved numerically.

As a result, students should learn both analytical and numerical solution techniques in their mathematics courses. They should understand the reasons for selecting a particular technique, develop an understanding of the range of applicability of the technique, acquire familiarity with the mechanics of the solution technique, and understand the limitations of the technique.

In order to acquire this level of understanding, analytical solution techniques should be taught in conjunction with numerical techniques. Analytical solutions should generally be introduced using relatively simple examples so that students can develop a sound understanding of the concepts. Numerical techniques for solving similar problems should also be introduced so that students can make the connection between the analytical and numerical solutions. For more complex problems, numerical solutions should be emphasized; however, analytical solutions (perhaps under constrained conditions) should be re-emphasized to stress the need for validating numerical solutions. With this approach students can develop a sound understanding of the fundamental concepts presented and learn solution techniques that can be applied to solving realistic problems.

The development of problem-solving skills is one of the primary goals of the civil engineering curriculum. Problem solving involves five basic components: recognize and define the problem; formulate the model and identify variables, knowns and unknowns; select an appropriate solution technique and develop appropriate equations; apply the solution technique (solve the problem); and validate the solution. Solution validation is one of the most important steps in this process and includes interpreting the solution, identifying its limitations, and assessing its reasonableness using appropriate approximate solutions or common sense.

As an example of this approach, consider the following problem. Design a cylindrical aluminum can that can hold 400 mL and that minimizes the quantity of aluminum used. The solution for this problem can be described as follows:

Problem Solving Component	Solution
1. Recognize and define the problem	Minimize the surface area of the can
2. Formulate the model and identify variables, knowns and unknowns	Surface Area = $2\pi R^2 + 2\pi Rh$ Volume = $\pi R^2 h = 400$ cm^3 Radius R and height h are unknown
3. Solution Technique	When the derivative of the surface area with respect to the radius equals 0, the surface area is a minimum or maximum. Differentiate analytically to find dA/dR.
4. Solve the equation	$dA/dR = 4\pi\text{R} - 800/R^2 = 0$ $R = 3.99$ cm and $h = 7.98$ cm
5. Validate the solution	Is the solution a minimum? Evaluate the 2nd derivative: $d^2A/dR^2 = 4\pi + 1600/R^3$ which is positive for R greater than zero.Or test the surface area equation using values of R that are greater and smaller than 3.99, and determine that $R{=}3.99$ provides the minimum surface area.

We believe that it is essential for students to be exposed to problem solving techniques in their mathematics courses as well as in their engineering courses. It is not necessary that the problems used in mathematics courses be related to engineering, but it is essential that students gain continual exposure to the tools required to solve engineering problems and to develop the ability to apply their mathematical skills to applied problems.

Technology

Technology (i.e., computer software, graphing calculators) must be a major component of mathematics curriculum. However, it is important that technology be applied in an appropriate manner and at the appropriate time in the curriculum. Many students entering college have been exposed to technology in their classrooms for several years and in their day-to-day lives. One of the keys to maintaining their enthusiasm for mathematics is to show them how technology can be used to solve real problems and to help them to realize that this same technology will be used in their engineering classes and in their careers.

The incorporation of technology into the mathematics curriculum means that some topics must be sacrificed. However, one of our recommendations for changing the mathematics curriculum is to place less emphasis on learning complex analytical solution techniques and to spend more time emphasizing the application of numerical solutions and computer based analytical techniques, problem analysis, and problem solving skills. By foregoing these more complicated analytical techniques two goals can be accomplished; more emphasis can be placed on gaining a deeper understanding of concepts through the application of both analytical and numerical solutions, and students gain an appreciation for the benefits of technology in mathematics and engineering.

Thus, while students need to apply technology to solve complex problems, the technology must be used in a manner that promotes understanding of mathematical concepts and their applications. In many instances, students learn how to apply technology (i.e., they know how to make software provide answers) without understanding the solution approach. This trap must be avoided.

Several basic technology skills should be taught in introductory courses to complement the theoretical and analytical treatment of mathematics topics. These address the areas of graphical techniques, algorithm development, symbolic manipulation and equation solving, spreadsheet operations, and basic computer programming. Students should be able to:

- Graph analytical functions using a graphing calculator or computer.

  Many students are visual learners and the ability to graph functions is an essential part of grasping new concepts and evaluating the effects of different parameters on the shape of a function.

- Use spreadsheet and data management software such as Microsoft Excel and Microsoft Access during their introductory courses.

  The level of familiarity should include the use of macros in spreadsheet programs and the capability to perform statistical analyses and equation solver routines.

- Use equation solvers and programs capable of symbolic manipulation such as MathCad.

- Implement a computer language such as Visual Basic.

  The goal in learning a computer language is to develop an understanding of programming techniques. The depth of knowledge need not be substantial and the selection of the specific language is less important than the development of detailed logic and flow charts. Students who go beyond an undergraduate degree will need programming skills to a much greater extent, but these skills can be acquired in more advanced courses. In addition, specific software packages are often taught within specialty courses.

With all of these tools it is essential that they be introduced to students at the appropriate time; after the student acquires a conceptual understanding of the technique but before the tedium of solving problems by hand frustrates them. In addition, the use of these tools should be coordinated between mathematics and engineering. The need for cooperation between faculty in these disciplines cannot be underestimated.

Ideally, implementation of this cooperative approach will take place over three to four years of students' undergraduate career. Indeed, in order for the mathematics content to link to the engineering curriculum content, it may be appropriate for students to take introductory mathematics courses in their freshman, sophomore and junior years.

Instructional Techniques

A number of instructional methods have proven effective for developing mathematical comprehension. The most important of these is the use of hands-on, active learning techniques in the classroom.

Of equal import is the need to make students understand the utility of the material they are being taught. Students need to understand and appreciate the need for their courses. Many engineering students leave their mathematics courses thinking that the material will never be used in their engineering courses. It is essential that mathematics courses have some future value in their engineering courses. The mathematics portion of a student's curriculum should not be simply something "to get through." This means that engineering and mathematics faculty must coordinate their curriculum. Mathematics faculty must teach methods that are applicable to current engineering practice, and engineering faculty must employ these methods in their curriculum within a reasonable time period after students learn the techniques.

Finally, it is necessary that institutions and national organizations provide resources for faculty development to implement curriculum reform and new teaching methods. Every Ph.D. program in mathematics and engineering should have a formal requirement for teacher development. Departments should provide incentives for current faculty to become involved in curriculum and pedagogical reform.

Instructional Interactions

Engineering pedagogy has advanced dramatically over the past decade. There has been a significant increase in team interaction. Our courses rely more heavily on the use of technology for instruction and as part of the curriculum. Active and cooperative learning is stressed in many of our classes. Many professors have reduced the number of topics in their classes in order to provide greater understanding and retention. There is a stronger emphasis on teaching problem-solving skills and providing more open-ended problems that closely simulate the types of problems encountered in practice.

While many of these topics have also been addressed in the mathematics curriculum, there have been very few examples where engineering faculty have been involved or even informed as to the types of changes that have been implemented. One of the most significant conclusions of this report is that there is a need for much greater interaction between engineering and mathematics faculty regarding curriculum reform. This interaction needs to occur formally at both the national and institutional levels. Existing forums such as the American Society of Engineering Education and the Mathematics Association of America can be used to initiate interactions at the national level. It is incumbent upon each university involved in mathematics and engineering reform to coordinate curriculum changes to ensure success. In addition, more studies are required that track students and assess the outcome of students who have been involved in curriculum reform.

WORKSHOP PARTICIPANTS

Robert Henry, Associate Professor and Associate Dean, Engineering and Physical Sciences, University of New Hampshire

Marc Hoit, Professor and Associate Dean, Civil Engineering, University of Florida

Lynn Katz, Associate Professor, Civil Engineering, University of Texas at Austin (coordinator)

Saleh Keshawarz, Associate Professor and Chair, Civil and Environmental Engineering, University of Hartford

Fred Hart, Professor and Head, Civil and Environmental Engineering, Worcester Polytechnic University

Matt Ohland, Assistant Professor, General Engineering, Clemson University.

Ben Sill, Alumni Professor and Director of General Engineering, Civil Engineering, Clemson University

David Thompson, Associate Professor, Civil Engineering, Texas Tech University

Mathematics Participants

Susan Ganter, Associate Professor, Mathematical Sciences, Clemson University

William McCallum, Professor, Mathematics, University of Arizona

APPENDIX: Mathematics Topics Relevant to Civil Engineering

Basics

- Algebra
- Trigonometry
- Logarithms
- Graphical Analysis of Functions
- Data Transformation (e.g., $y = axb \Rightarrow \log y = \log a + b\log x$)
- Systems of Equations
- Iterative Solutions

Vectors

- Addition
- Dot Products
- 3-D Coordinate Systems
- 3-D Visualization
- Coordinate Transformation
- Translation, Rotation

Matrix Algebra

- Linear Systems
- Transformations
- Determinants (3x3)

Series

- Taylor Series (Approximations)

Differentiation

- Definition of a Derivative
- Derivative as a Slope of a Curve
- Min, Max
- Derivatives of Polynomials, Trigonometric Functions and Exponentials
- Chain Rule
- Root Finding

Integration

- Definite vs. Indefinite Integrals
- Integration as the Area under a Curve
- Basic Integration Techniques
- Multiple Integrals (Two)
- Areas, Volumes, Centroids, Moments
- Numerical Integration

Differential Equations

- First Order ODE – Linear, Constant Coefficient
- Second Order ODE – Single Variable
- Initial and Boundary Conditions
- Laplace Transformations

Partial Differential Equations (Intro. – Single Spatial Variable)
Numerical Methods – Euler's, Runga Kutta

Probability/Statistics/Optimization

Regression, Curve Fitting
Distributions — Triangular, Normal
Return Intervals
Linear Programming — Simplex Method

Engineering:
Electrical Engineering

CRAFTY Curriculum Foundations Project
Clemson University, May 4–7, 2000

Ben Oni, Report Editor
Kenneth Roby and Susan Ganter, Workshop Organizers

Summary

This report focuses on establishing the foundation mathematics needed to support the study and practice of electrical engineering with emphasis on the undergraduate level.

To strengthen communication between communities of mathematicians and electrical engineers, we have prepared this document to highlight the areas of mathematics that are most applicable to the study and practice of electrical engineering.

For **outcome objectives**, we propose that the mathematics taught to undergraduate electrical engineering students should help them in developing skills to:

1. Formulate problems in electrical engineering from real life situations,
2. Conceptualize the outcomes of electrical problems,
3. Simplify complex problems and estimate the reasonableness of solutions,
4. Visualize solutions graphically from inspection of their mathematical descriptions,
5. Visualize the form of a time function by inspection of the poles and zeros of its frequency transform,
6. Be able to mathematically model physical reality,
7. Perform rudimentary analysis in electrical engineering,
8. Validate solutions to electrical engineering problems.

Narrative

Introduction and Background

Electrical engineering deals with the manipulation of electrons and photons to produce products that benefit humanity. The design of these products is based on scientific principles and theories that are best described mathematically. Mathematics is thus the universal language of electrical engineering science.

Undergraduate electrical engineering education must provide students with the conceptual skills to formulate, develop, solve, evaluate and validate physical systems. Our students must understand various problem-solving techniques and know the appropriate techniques to apply to a wide assortment of problems. We believe that the mathematics required to enable students to achieve these skills should emphasize concepts and problem-solving skills more than emphasizing repetitive mechanics of solving routine problems. Students must learn the basic mechanics of mathematics, but care must be taken that these mechanics do not become the primary focus of any mathematics course.

More generally, it is vitally important that electrical engineering students recognize the importance and beauty of mathematics in their chosen profession. We feel strongly that students will appreciate the power of mathematics if each mathematics course clearly states its objectives at the outset. Students should be told *what* they are going to study, *why* they are going to study it, and *how* it fits into the engineering profession. This motivation will need to be repeated throughout each course.

Many undergraduate mathematics curricula currently supporting electrical engineering programs could be modified to better meet the needs of these programs. What follows are common weaknesses (from the viewpoint of electrical engineering) seen in many mathematics curricula.

1. Too much time and emphasis are placed on topics that are not widely used while topics that have widespread use often receive cursory treatment. One example is the excessive time and attention spent on various solution techniques for ordinary differential equations. Although understanding the structure of solutions for first- and second-order, constant coefficient differential equations is important for electrical engineering problems, more useful and widely used are Laplace transforms and related techniques. Yet these latter topics are often given cursory treatment in favor of more general structure theory.
2. There is often a disconnect between the knowledge that students gain in mathematics courses and their ability to apply such knowledge in engineering situations. Perhaps, the use of more engineering or real life examples will reduce this disconnect. Based on current learning theory, efforts to focus on underlying principles (not necessarily abstract statements of mathematical concepts) that are applicable in many different contexts are effective in helping students to transfer knowledge.
3. Current mathematics curricula for engineering are front-end loaded. Consequently, as a matter of timing, many topics are presented too early and cannot be reinforced soon enough through engineering applications before students forget the topics
4. Too often, mathematics is taught as a list of procedures or as theorem-proof exercises without grounding the mathematics in reality. While we do not expect mathematics instructors to be well versed in all engineering applications, we would like examples of mathematical techniques explained in terms of the reality they represent. We strongly urge that team taught mathematics courses be considered. Teams would consist of mathematics and electrical engineering professors. We feel that team-teaching could better motivate and enthuse our students.
5. Failure to utilize appropriate technological tools while continuing to focus on mastery of symbolic manipulation often encourages memorization and rote algorithm practice at the expense of conceptual and graphical comprehension. Introducing symbolic manipulation programs, e.g., *MathCAD*, *Mathematica*, *Maple*, would be valuable to subsequent electrical engineering courses whose instructors choose to allow/encourage students to perform routine symbolic and numerical manipulations using such programs.
6. The first two years of mathematics that support instruction in electrical engineering should present students with conceptual understanding of mathematical disciplines other than just single variable calculus, multivariable calculus and ordinary differential equations. Other mathematical subjects that are important for electrical engineering students include linear algebra, probability and stochastic processes, statistics, and discrete mathematics.

Electrical engineering is an exciting and creative profession. Those engineers possessing an understanding and facility of mathematics have an opportunity to be among the most creative of designers. Students need to know and to feel how important, how useful, and how meaningful mathematics is. Many courses stress the drudgery, not the beauty. This needs to be changed.

Electrical Engineering Subdisciplines

To describe our mathematics recommendations in sufficient detail, the undergraduate electrical engineering curriculum is broken down into the following broad areas:

1. Electrical Circuits
2. Electromagnetics
3. Systems, including Controls, Linear and nonlinear Circuits, and Power
4. Signals
5. Design
6. Microprocessor/Computer Engineering

What follows are summaries of the proposed mathematics requirements for each subdiscipline.

1. Electrical Circuits

The electrical circuits course is the passageway to electrical engineering. Of critical interest are the logical thinking skills to analyze electric circuits. In this course, students are introduced to the application of physical laws, e.g., Ohm's, Faraday's and Kirchoff's, in electrical engineering. Students are also introduced to the electrical engineering foundation elements: resistor, inductor and capacitor, and their response *(voltage, current and power profiles)* to DC, steady state AC, and transient stimuli respectively. In most institutions, the circuits course is a two-part series with DC circuit analyses and transient response offered in the first semester and AC circuits and steady state response offered the second semester.

A. DC Circuits. Typical problems in this section involve the simplification of series, parallel and mesh circuits. Analyses of these circuits require setting up, manipulating, and obtaining solutions to algebraic equations. Subtle mathematical skills in the understanding of the circuit problems also include direct and inverse proportionality to enable students to understand voltage and current divider rules respectively.

In the circuit areas dealing with power, and power transfer, knowledge of integral calculus and basic differentiation is required especially for maximum power transfer analysis.

B. AC Circuits. This part of circuit analysis deals with the response of different circuit configurations and elements to steady state sinusoidal inputs. Different mathematical techniques are necessary to simplify the circuits before gaining understanding of the response. The foundation mathematics necessary for the analyses include:

Concept of functions, especially sinusoidal functions. Students need to understand and visualize profiles of basic functions. Use of common real life examples is strongly suggested in teaching this topic.

Application of trigonometric identities to sinusoidal analyses.

Manipulation and representation of sinusoidal functions in Euler, polar, and rectangular coordinates.

Complex algebra.

C. Transients. This topic deals with response of discrete circuit elements, or combinations thereof, to electrical stimuli. At the DC stage, the typical stimulus is the step. The mathematical background required in the analysis includes exponential functions and introductory differential equations. In the latter subject the primary focus should be on ***standard solutions*** to first and second order differential equations with constant coefficients rather than on more general techniques for solving differential equations. The preferred and more useful approach to solving differential equations in electrical engineering is via the Laplace transform. Laplace transform methods reduce differential equation problems to algebraic formats with which students feel more comfortable. This topic should be presented during the first year of undergraduate mathematics.

At the AC stage, the typical stimuli are sinusoidal, triangular and square functions. Usually, the interest in this setting is steady state rather than transient. The Laplace transform still provides the preferred method of analysis because, in addition to reducing differential equation problems to algebraic equation problems, it also incorporates initial conditions in the solution.

2. Electromagnetics

Study of electromagnetic fields and waves is a crucial area in electrical engineering for which understanding of vector algebra and vector calculus is required. The basic laws of electromagnetics are summarized in Maxwell's equations:

$$\text{Faraday's Law: } \vec{\nabla} \times \vec{E} = \frac{\partial \vec{B}}{\partial t}$$

$$\text{Ampere's Law: } \vec{\nabla} \times \vec{H} = \vec{J} + \frac{\partial \vec{D}}{\partial t}$$

$$\text{Gauss' Law: } \vec{\nabla} \cdot \vec{D} = \rho$$

$$\text{No Isolated Magnetic Charge: } \vec{\nabla} \cdot \vec{B} = 0$$

These are partial differential equations that require deep conceptual understanding of vector fields and operations related to vector fields: gradient, divergence, and curl. With the increasing power and availability of software, e.g., *Mathematica*, *Maple*, and *Matlab*, to perform the actual manipulations, it is crucial that students develop a conceptual understanding of vector fields and related operations.

It is less important to emphasize the actual manipulations. For example, it is less important that a student be given a scalar-valued function of several variables and be asked to compute the gradient. It is more important that students be able to start with a contour plot (topographic map) of a scalar-valued function of several variables and draw the gradient function.

It is less important that students start with a vector field and be able to compute the divergence or curl. It is more important that students be able to interpret verbally and graphically pictures of vector fields.

Students should be able to identify regions in which the magnitudes of the divergence or curl will be large or small. To support conceptual understanding, graphical interpretation, and verbal description it is helpful to connect students of vector calculus with applications such as electromagnetic fields, fluid mechanics and heat transfer.

The study of electromagnetics requires a conceptual understanding of partial differential equations and their solutions, and the power and limitations of numerical solutions techniques. The study of specific partial differential equations that permit closed-form solutions is less important than the development of this conceptual picture.

Since most students in electrical engineering do not begin studying electromagnetic fields and waves until their junior year, it is important that the relevant topics of vector calculus and partial differential equations not be taught before the second semester of the sophomore year. Timing of the topics is important to help students connect their studies in mathematics with their study of electromagnetics. Individual schools should encourage conversations between faculty in electrical engineering and mathematics to prepare a mathematics curriculum that is responsive to the specific requirements of the electrical engineering department.

3. Systems

Control Systems—Linear and NonLinear Circuits. One purpose of systems analysis is to represent reality mathematically. At the undergraduate level, linear time-invariant systems are discussed, studied and designed. The systems may be continuous or discrete and may have one or more inputs and one or more outputs. The system is modeled as a "box," a device that modifies the signals entering it resulting in an output according to the transfer function of the system:

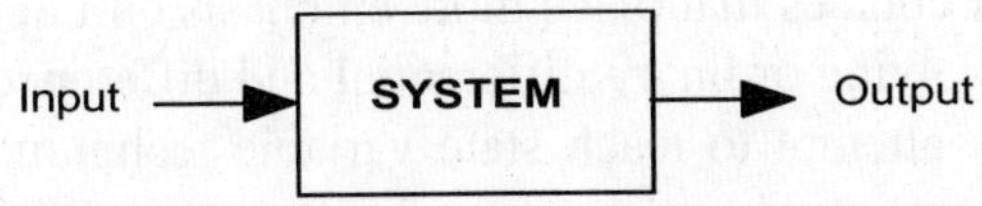

Within electrical engineering, the systems problem has the following forms:

1. Find the transfer function of a SISO (single input-single output) system,
2. Find the transfer function of a MIMO (multi input-multi output) system,
3. Given the transfer function of a SISO system, what is the output of the system if the input to the system is a specified function?
4. Given the transfer function of a SISO system, how must the system be modified to satisfy given specifications?
5. Given the state equation matrices for a MIMO system, what are the outputs of the system if the input to the system is a specified vector function?
6. Given the state equation matrices for a MIMO system, how must the system be modified to satisfy given specifications?

In undergraduate courses the systems studied are linear and time-invariant. Continuous systems can be modeled by ordinary differential equations although the order of these equations might be quite high. Software packages such as MATLAB are used extensively in most systems courses.

For **continuous systems** the mathematical tools needed consist of:

1. Laplace transforms and techniques such as partial fraction expansions and residues
2. State variable techniques including eigenvalues/eigenvectors, interpretation of the matrices, etc.
3. Basic differential equations, focused on standard solutions to common problems.

For **discrete systems** difference equations are used instead of differential equations and the discrete-time state model is used. The mathematical tools needed are:

1. Difference equations
2. The state transition matrices and solutions to discrete-time state models
3. Z-transforms
4. Discrete Fourier transforms
5. Fourier analysis.

For both continuous and discrete systems it is important to be able to use the poles and zeros of transformed time functions to visualize the system's time response to various inputs. For continuous systems the s-plane is important. Students should be able to plot the poles and zeros of the transfer function and from this plot know the form of the impulse response by inspection. They should understand how the locations of the poles affect the output of the system. They should see how the locations of zeros affect the various modes of the system. In other words, they need to see the time response in the s-plane and understand the physical realities encoded in the poles and zeros. For discrete systems the same holds true for the z-plane.

The mathematical courses supporting systems are primarily linear algebra and ordinary differential equations. As stated above, ordinary differential equations courses tend to overemphasize the development of numerous solution methods for first- and second-order differential equations. In truth, systems deal with higher order differential equations, and for SISO continuous systems Laplace transforms are almost always the preferred method of solution. Discrete systems use the method of z-transforms. Both continuous and discrete MIMO systems use state variable techniques.

Consequently, we would prefer courses that place more emphasis on Laplace transforms, z-transforms, and state variable techniques for solving ordinary differential and difference equations.

Linear algebra courses usually attempt to teach state variable techniques. We recommend that these courses further develop the concepts taught in the "new" differential/difference equations course. State transition matrices and their properties could be studied. Eigenvalues and eigenvectors could be explained. The relation between the roots of the characteristic equation and eigenvalues could be stated. Other topics might include:

1. Techniques for computing the matrix exponential and its integral,
2. Eigenvalue-eigenvector methods for computing matrix exponentials,
3. Decomposition of time-invariant state matrices,
4. Jordan forms,
5. Singular value decompositions and state space applications.

The elements of linear systems could be taught during the sophomore year.

Power Systems comprise the study of the transmission and distribution of electric power. The study of power systems depends upon a firm mathematical grounding in the use and manipulation of trigonometric functions as well as algebraic manipulation of complex numbers. The use of phasor notation (an application of polar co-ordinates) plays a central role in power systems analysis. Students must also know Euler's formula and be facile in going from polar to rectangular co-ordinates and vice-versa.

Power systems analysis requires not only algebraic manipulation but also recognition of the changes a signal undergoes and the form of the signal that results. While a solution having the form

$$f(t) = 5e^{-2t} - 5e^{-3t}(\cos t + \sin t) \quad \text{volts},$$

may be correct, the form

$$f(t) = 5e^{-2t} - 5e^{-3t}[1.414\cos(t - 45°)] \quad \text{volts}$$

is more useful. The student can visualize a sinusoidal wave with an amplitude of 1.414 volts that lags the input signal by 45. Next, the student can visualize a sinusoid that decays exponentially. Thus, the waveform with its phase angle can be easily visualized, whereas the sum of sinusoids gives little information about the phase angle. Power systems may be taught as early as the student's 5th semester of undergraduate studies.

4. Signals/Communications

One of the most fundamental applications in electrical engineering is the transmission, modification and reception of signals. Communication systems is concerned with:

1. The transmission of signals through electric networks
2. The modulation and demodulation of signals
3. Sampling
4. Noise
5. Statistical methods of information transmission systems

Digital signal processing is an important area within electrical engineering. The digitization, modulation, transmission, demodulation, and reception of signals is vital to modern communications. Image processing and pattern recognition techniques fall within the purview of digital signal processing.

Communications and digital signal processing are taught in depth usually during the last two or three semesters of the student's undergraduate studies. The understanding of mathematical concepts is essential within the communications area. Of particular importance are:

1. Basic algebraic techniques
2. Basic trigonometric identities
3. Integration techniques, including partial fractions and integration by parts
4. Taylor Series Expansion (i.e., linear approximation, expansion out to two or three terms)
5. The Fourier Transform and its use
6. Fourier Series
7. The use of the Laplace Transform
8. The use of the Z Transform
9. Probability and Stochastic Processes

5. Design

Design and modeling are two generic tasks in which engineers participate after completing their undergraduate degrees. In addition to preparation in mathematics for the other disciplines within electrical engineering, there are additional areas in mathematics that are necessary to support learning and growth in design and modeling. Such areas include statistics, empirical modeling, parameter estimation, system identification, model validation and design of experiments. Demand from industry for expertise in these areas appears to be much stronger than demand within electrical engineering curricula. That may explain why these areas are not prerequisites for courses in electrical engineering. However, expertise in these areas is increasingly important for electrical engineering graduates.

Topics from these areas that will be valuable for engineering graduates include the concept of a random variable, analysis of sets of data, concepts of sample means, sample variances and other sample statistics as random variables, and hypothesis testing. To illustrate why these topics are important here are some examples of applications.

First Example, Simple Parameter Estimation: Construct a circuit containing a resistor (resistance = R) and a capacitor (capacitance = C). If the capacitor is initially charged and then discharged through the circuit, voltages and currents decay exponentially. Data on a particular voltage can be taken at various points in time. Students then must estimate the time constant (= RC) using the accumulated data. There are a variety of techniques through which estimates of the time constant can be obtained. Students need to be familiar with the techniques as well as the supporting concepts and broader applications.

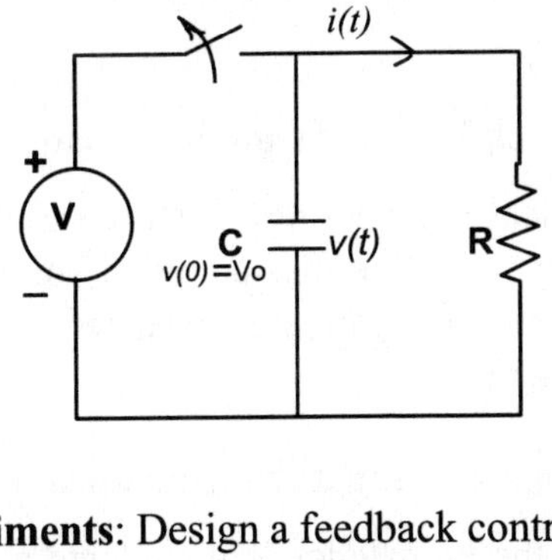

Second Example, Design of Experiments: Design a feedback controller that meets several specifications and minimizes percent overshoot. There are a number of parameters that may be adjusted. Students should be able to design a set of experiments that will help determine narrow intervals for the parameter values in order to optimize the design.

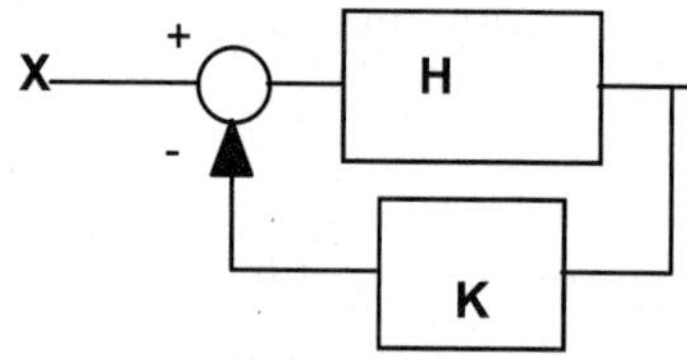

Third Example, Model Validation: Develop an empirical model for a complex physical process. Once the model is produced, students should be able to develop a set of experiments to help them understand the validity of the model.

6. Computer Engineering/Microprocessors

Digital logic design and microprocessors require a mathematical background that is fundamentally different than the background necessary for the areas discussed above. Circuits, electromagnetics, signals, and systems require mathematics in which the variables can be any real number, i.e., *continuous mathematics*. Digital logic design and microprocessors require mathematics in which the variables can only assume values in a finite set, so-called *discrete mathematics*. Students need instruction that emphasizes the fundamental difference between continuous and discrete mathematics.

More specifically, students need Boolean algebra and finite state systems. For Boolean algebra, they need to understand truth tables for the basic operators: NOT, AND, OR, NAND, and NOR. They need to analyze combinational networks constructed from these basic operators and methods by which the networks may be simplified. Examples that help students relate combinational networks to actual applications will help build motivation and understanding.

For finite state systems, students need to understand the concepts of a finite state machine and a state transition diagram. Understanding these concepts can be strengthened by examining Mealy and Moore realizations and the equivalence between the two realizations. In addition, connections between finite state machines and regular expressions should be explored. Finally, students need to start with a description of a physical situation, synthesize a state transition diagram, and then design the combinational logic that together with memory can realize the state transition diagram.

Understanding of these concepts from Boolean algebra and finite state machines will provide students with the necessary mathematical background to study computer engineering and microprocessors.

Understanding and Content

What follows are brief summaries of our responses to the specific questions posed by the Curriculum Foundations Organizing Committee.

What conceptual mathematical principles must students master in the first two years?

The mathematics required for electrical engineering students should emphasize concepts and problem solving skills more than emphasizing repetitive mechanics of solving routine problems. Students must learn the basic mechanics of mathematics, but care must be taken that these mechanics do not become the primary focus of any mathematics course.

What mathematical problem solving skills must students master in the first two years?

There is often a disconnect between the knowledge that students gain in mathematics courses and their ability to apply such knowledge in engineering situations. Perhaps, the use of more engineering or real life examples will reduce this disconnect. Too often mathematics is taught as a list of procedures or as theorem-proof exercises without grounding the mathematics in reality. While we do not expect mathematics instructors to be well versed in all engineering applications, we would like examples of mathematical techniques explained in terms of the reality they represent.

Students of electrical engineering need to be skillful at mathematically modeling physical reality. They need to be able to simplify complex problems, estimate the reasonableness of solutions, and visualize solutions graphically from inspection of the mathematical descriptions.

What broad mathematical topics must students master in the first two years?

The first two years of mathematics that support instruction in electrical engineering should present students with conceptual understanding of mathematical disciplines other than just single variable calculus, multivariable calculus and ordinary differential equations. Other mathematical subjects that are important for electrical engineering students include linear algebra, probability and stochastic processes, statistics, and discrete mathematics.

Listed below are the most important mathematical topics that we believe students in electrical engineering should learn during the first two years of undergraduate studies. All of these topics were discussed earlier in this report. The Appendix provides another summary of topics, this one organized by the six sub disciplines of electrical engineering that were identified in the previous section.

- Manipulation, solution, and analysis of real and complex algebraic equations
- Basic differential and integral calculus
- Standard solutions for basic differential equations, in particular first- and second-order differential equations with constant coefficients
- Laplace, Fourier and Z transforms
- Vector calculus
- Taylor series
- State variables and finite state systems
- Difference equations
- Probability and stochastic processes
- Statistics
- Model validation
- Parameter estimation—techniques and application
- Boolean algebra—analysis and application

Technology

How does technology affect what mathematics should be learned in the first two years?

New engineering and mathematical software only reduce the dependency on routine, excessive and repetitive mathematical computations. Software should not be used to replace the necessity to teach students how to pose and formulate mathematical questions and how to evaluate answers obtained for these questions.

What mathematical technology skills should students master in the first two years?

The use of mathematical software is necessary. In this regard, mathematics departments are strongly encouraged to routinely use common math software tools and to promote students' use of them. Failure to utilize appropriate technological tools while continuing to focus on mastery of symbolic manipulation often encourages memorization and rote algorithm practice at the expense of conceptual and graphical comprehension.

What different mathematical technology skills are required of different student populations?

We did not identify any difference in requirements.

Instructional Techniques

What are the effects of different instructional methods in mathematics on students in your discipline?

We are not aware of differing effects.

What instructional methods best develop the mathematical comprehension needed for your discipline?

Hands-on, interactive engagements and project-based learning methods have been observed to promote students' learning.

Team teaching of mathematics courses should be considered. Teams could consist of faculty members from both mathematics and electrical engineering. We believe that team-teaching could better motivate and enthuse electrical engineering students.

Each mathematics course should clearly state its objectives. Students should be told *what* they are going to study, *why* they are going to study it, and *how* it fits into the engineering profession.

What guidance does educational research provide concerning mathematical training in your discipline?

We are not aware of the relevant educational research.

Instructional Interconnections

What impact does mathematics education reform have on instruction in your disciplines?

We are not aware of or familiar with current directions in mathematics education reform

How does education reform in your discipline affect mathematics instruction?

We did not have a response to this question.

How can dialogue on education issues between discipline and mathematics best be maintained?

One effective method for continuing this dialogue between mathematics and electrical engineering is to have more extensive contact between individuals in each discipline and the major professional associations in each discipline. Professional engineering associations which should be contacted include:

The American Society for Engineering Educators (ASEE)

The National Electrical Engineering Department Heads Association (NEEDHA)

Similar organizations for mechanical, civil and chemical engineering

WORKSHOP PARTICIPANTS

Ben Oni, Associate Professor of Electrical Engineering and Acting Dean of Engineering, Tuskegee University

Brian Butz, Professor of Electrical Engineering, Temple University

Jeffrey Froyd, Research Professor in Electrical Engineering, Dwight Look College of Engineering, Texas A&M University

Ronald Talley, Department Head and Professor of Electronic and Engineering Technolgy, Tri-County Technical College, South Carolina

Mathematics Participants

William Barker, Professor of Mathematics, Bowdoin College

Robert Fennell, Professor of Mathematical Sciences, Clemson University

APPENDIX: Specific Issues on Understanding and Content

1. Electrical Circuits

Algebraic equations, manipulations and solutions

Differential and integral calculus

Concept of functions

Exponential functions

Sinusoidal functions, representation and manipulation in Euler, polar, rectangular coordinates

Application of trigonometric identities

Algebra of complex numbers

Introductory differential equations - focus on *standard solutions* to problems with basic inputs including step, sinusoids, triangular, square functions *(teach by end of first year)*

Laplace transform (emphasize this topic over differential equation techniques in the first year)

2. Electromagnetics

Vector calculus (Do not teach before the 2nd semester of the sophomore year)

Conceptualization

Operation

Gradients

Divergence

3. Systems

Continuous Systems

Laplace transforms and techniques

Integration by parts

Partial fraction methods

State variables

Eigenvalues and eigenvectors

Basic differential equations—*focus on standard solutions to common problems*

Discrete Systems

Difference equations

Systems of first order differential equations

Z-transforms

Discrete Fourier transform

Fourier analysis and techniques

Power Systems

Sinusoidal functions

Algebra of complex numbers

4. Signals/Communications

Fourier analysis, transform and techniques
Integration by parts
Partial fraction methods
Probability and stochastic processes
Z-transform
Taylor series (linear approximation—interested only in the first two terms)

5. Design

- Statistics
  - Data collection
  - Sampling
  - Analysis (distribution, graphical techniques)
- Concepts of random variables
- Model validation
- Parameter estimation
- System identification

6. Computer Engineering/Microprocessors

Boolean algebra
Finite state systems

Engineering: Mechanical Engineering

CRAFTY Curriculum Foundations Project
Clemson University, May 4–7, 2000

David Bigio, Report Editor
Kenneth Roby and Susan Ganter, Workshop Organizers

Summary

We have endeavored to answer the questions as posed in order to satisfy the understood needs of the Mathematical Association of America. However, the answers to just the posed questions do not address the bigger picture that was the underlying theme of our discussions. We therefore present an overall context of what we want the students to "be" and then a list of the details within each mathematics topic.

We want students who are able to:

- Learn in context
- Conceptualize
- Set up equations
- Apply problem solving techniques
- Translate mathematical results
- Understand the language of mathematics vs. language of engineering (in all dialects)
- Use modeling techniques
- Understand logic discipline
- Manipulate complex equations

We want students to understand the physics and how equations describe the physics. We want them to then solve problems with whatever tools are appropriate *and* to understand the solutions.

Narrative

Introduction and Background

We have tried to address our answers to the MAA questions with respect to where we think the Mechanical Engineering curriculum is now and where it is going to be in the near future. The different themes we think are important included:

- "Slide rule engineering" and packaged knowledge is insufficient for modern problems
- Smaller scale systems
- Nanoscale-thin layers

- Smarter systems—more intelligent (Mechatronics)
- Modularization
- Precision movement
- New materials—soft/wet, increasingly complex (using tensor notation)
- Biological and medical engineering
- Analytic complexity and interrelatedness of fields
- Three types of systems
  - Static—large scale stresses
  - Dynamic—phenomenon over time
  - Non-deterministic—statistics and probability
- Increasing use of discrete techniques in addition to continuous
- Central roles of calculators and computers
- Massive codes used for problem solving
- Need to understand microprocessors and mechanical systems—Boolean logic

Regardless of what the new foci will be one thing is clear: the way students need to know the material is different than before. Students need to be able to deal in a more complex interface with the different disciplines. They need to understand how to take the fundamentals and use them in different ways. Calculators, computers, and packaged software give students a facility to see the physical reality in a variety of ways.

Understanding and Content

We believe that the basic mathematical foundations have not changed a great deal. It is, however, the implementation of this knowledge and the drive for pedagogical transformation that is of concern in the future. The following are the group's responses to the specific MAA questions, organized as requested by considering different aspects of what is desired in the first two years of undergraduate instruction.

What conceptual mathematical principles must students master in the first two years?

The following table states the specific mathematical principles and where they are used in the mechanical engineering curriculum.

Mathematical Concept	Context for Mechanical Engineering
Algebra and trigonometry Coordinate systems	All areas of mechanical engineering
Geometry Transformation: rotation & translation	All areas of mechanical engineering Notation facility Codes
Estimation Order of magnitude Distinguishing realistic from nonsense	Thermo fluids Engineering design Physical design of experiment Modeling
Statistics Mean and standard deviation	Experimental design and analysis Design Quality Assurance Manufacturing

Mathematical Concept	Context for Mechanical Engineering
Derivatives and integrals	Statics and mechanics Thermodynamics Energy and work Control theory
Taylor series and Fourier series	Vibrations/Instrumentation System Dynamics
Ordinary differential equations: Linear first and second order Homogeneous and non-homogeneous	Heat conduction Fully developed flow Solid state diffusion Dynamic systems
Multivariable calculus Gradients Surfaces and planes	Fluids Conduction
Linear Algebra Vectors and matrices Eigenvalues and eigenvectors	Dynamics and systems dynamics Vibrations Controls Nonlinear finite element analysis Optimization
Vector analysis	Statics, dynamics, and kinematics Mechanics of materials
Partial differential equations	Thermal fluids Soils mechanics
Numerical Methods Discretization Use of software such as *Matlab*	Dynamics Dynamical systems Vibration Solid and fluid mechanics
Complex variables	Controls (bode plots) Vibrations Electronic Circuit Theory

What mathematical problem-solving skills must students master in the first two years?

In an engineering discipline problem solving essentially means **mathematical modeling**: the ability to take a physical problem, express it in mathematical terms, solve the equations, and then interpret the result.

We feel that, especially in the first two years, the students are more comfortable and adept at being able to understand concepts through tackling sample problems. Part of the problem solving must move through Bloom's Taxonomy from mechanics to conceptualization to integration. The final stage could be done by moving some of the mathematics topics into the third year and coordinating the timing of the topics with the engineering program.

What broad mathematical topics must students master in the first two years? What priorities exist between these topics?

We provide a high level of specificity and detail in our presentation of the mathematics topics we consider important to an undergraduate in mechanical engineering. Moreover, rather than limit ourselves to just the first two years, we address all the mathematics skills we believe are necessary for our students, separating them into the three major levels of their educational career:

Secondary education—Major topics for prospective engineering students in high school include:

- algebra
- trigonometry
- geometry

A student should develop skills in these areas prior to entering an engineering program.

First two years of undergraduate education – Major topics to include are:

- geometry
- functions and graphs
- statistics
- integration and differentiation
- linear algebra
- ordinary differential equations
- partial differential equations
- numerical methods
- complex numbers and functions

Third Year—Advanced topics or topics focused on certain sub disciplines rather than across the curriculum are placed in this category.

The Appendix lists specific topics for each of these three educational levels. Moreover, instead of simply listing together the first two undergraduate years we suggest what the students should learn during the first year and what they should learn during the second.

What is the desired balance between theoretical understanding and computational skill? How is this balance achieved?

Students need enough of a conceptual foundation—not based on formal derivations—so that they get a basic understanding of the mathematical principles. They need to tie these with computational skills. Then computing will allow an effective application of the mathematics, yielding an understanding of the physical implications of the system under study. The emphasis is on physical understanding without the axiomatic structure.

What are the mathematical needs of different student populations and how can they be fulfilled?

We differentiate between students who terminate with a bachelor's degree and those who continue on for advanced degrees. The more theoretical mathematics courses should be made available during the last two years, when a student knows if they wish to go to graduate school.

Technology

How does technology affect what mathematics should be learned in the first two years?

Computers and appropriate software allow work on more complex problems earlier in a student's educational career. They also allow for visualization of the effect of varying parameters.

The internet gives access to vast amounts of information, allowing solution of more interesting complex problems.Many crank turning skills—like integration by parts, complicated integration substitutions, and manual manipulative skills—can be handled by technology. Instructors can thus spend more time teaching conceptual understanding of the important skills.

What mathematical technology skills should students master in the first two years?

Students need to be familiar with software applications for numerical computation and symbolic computation. They also need to be introduced to spreadsheets.

Instructional Techniques

What are the effects of different instructional methods in mathematics on students in our discipline?

Mathematics is the entry point for engineering students. We get our students because they "were good in math," and often because some instructor told them so. An ineffective mathematics program will drive our students away. The students can be turned off by poor instruction.

What instructional methods best develop the mathematical comprehension needed for your discipline?

Instruction is not effective when material is presented in an overly theoretical way, divorced from application. Topics should be presented in the context of physical concepts. Interdisciplinary team teaching/active learning/collaborative learning should be utilized as effective instructional methods. The use of student teams enhances instructional effectiveness.

Is it pedagogically more effective to teach mathematics to engineering students as a homogeneous group or as part of heterogeneous groups with students in other disciplines? We recommend that the MAA seriously consider which method of student grouping is the most conducive to student learning.

What guidance does educational research provide concerning the mathematical training in your discipline?

Several educational research insights are valuable in guiding mathematical training in our discipline. Specific examples include teaming and Bloom's taxonomy. Active learning has become an important activity for our students, both as preparation for the job market and as an effective way to deal with open-ended problems.

We propose that mathematics departments develop small projects that require teaming and active learning. This would help students learn fundamental mathematical principles, avoiding the mere memorization of algorithms that can arise from over-concentration on examples. There could be one or two projects during a semester, handled by teams of two to four students. Examples of subjects could be differential equations whose complexity increases from project to project. The applications could cut across different engineering disciplines so that students could develop a sense of the nature of each discipline.

Instructional Interconnections

What impact does mathematics education reform have on instruction in your discipline?

Some workshop participants felt that students are being trained in symbolic manipulation and memorization in ways that negatively impact their ability to attack more complicated problems. Our students seem to have a decreased ability to conceptualize. They are too dependent on methodology and not on principles.

How should education reform in your discipline affect mathematics instruction?

Educational reform in engineering, being driven by ABET. EC2000 has added more active learning, problem-based, team learning experiences. This results in more open ended, complicated problems for students to tackle. The paradigms for teaching in engineering are evolving from solely derivation and lectures to more varied methods, addressing a diversity of student learning styles. If mathematics is still taught in ways that do not address differing learning styles or teaching pedagogies, it will lose its effectiveness in preparing engineering students.

How can dialogue on educational issues between your discipline and mathematics best be maintained?

- Establish sites that implement change in the education paradigm.
- Organize workshops with mathematics and engineering faculty to advance this discussion.
- Establish joint meetings or sessions between MAA and FIE/ASEE.
- Recognize and encourage interdisciplinary cooperation via changes in the merit system.
- Encourage engineers to learn more about reforms under consideration by the MAA.
- Encourage books to be written jointly by mathematicians and engineers.

Funding needs to be found for such activities. The NSF is one appropriate funding source.

WORKSHOP PARTICIPANTS

Andrew Alleyne, Mechanical & Industrial Engineering, University of Illinois
William Beston, Engineering Science, Broome Community College
David I. Bigio, Mechanical Engineering, University of Maryland
P. K. Imbrie, Freshman Engineering, Purdue University
Phillip L. Jones, Mechanical Engineering and Materials Science, Duke University
Deborah A. Kaminski, Mechanical Engineering, Rensselaer Polytechnic Institute
Raymond L. Laoulache, Mechanical Engineering, University of Massachusetts, Dartmouth
Jed Lyons, Mechanical Engineering, University of South Carolina
Kenneth Roby, Mechanical Engineering, Clemson University (Deceased)
Dhushy Sathianathan, Engineering Design, The Pennsylvania State University
Robert L. Spilker, Biomedical Engineering, Rensselaer Polytechnic Institute

Mathematics Participants

David Bressoud, Mathematics, Macalester College
Harvey Keynes, Mathematics, University of Minnesota
John Peterson, Mathematics, Chattanooga State Technical and Community College

APPENDIX: Specific Mathematical Topics

Secondary School

- **Algebra/Trigonometry**

  Manipulation skills

  Logarithmic and exponential functions

  Trigonometric functions:

    In terms of right triangles and general triangles

    Expressing periodic phenomena

  Understanding variable names and use of symbols

- **Geometry**

  Sketching process for visualization

  Planar objects

First Two Years of Undergraduate Mathematics

- **Geometry**

  3D visualization (1st yr)

  Coordinate systems (1st yr)

  Transformations (1st yr)

    Translations

    Rotations

- **Functions and Graphs**

  Plotting—Cartesian, semi-log, log-log, (1st yr)

  Special functions and their properties—exponentials, step functions (1st yr)

  Parametric graphing (2nd)

- **Statistics**

  Mean/Standard Deviation/Variance (2nd yr)

  Regression Analysis (2nd yr)

    Least Squares

    Linear and Polynomial Regression Analysis

    Multivariable Regression Analysis (nonlinear)

  Probability—(2nd yr)

  Distributions—Gaussian, Poisson, Exponential (2nd yr)

  Examples:

  - Linear regression is used to find the slope of a stress-strain curve (the elastic modulus of a material).
  - Polynomial regression is used to model the rise and fall of hardness with aging time in precipitation hardenable aluminum alloys and predict the peak hardness conditions.
  - ANOVA and/or T-Tests are used to determine if there is a significant correlation between all kinds of data, such as heat treatment conditions and hardness.

- Basic statistics like average and standard deviation are used to describe the average grain size from metallographic measurements.
- The Weibull distribution is used as a model for lifetime and durability phenomena.
- The Exponential distribution is used as a model for reliability data.
- The Binomial distribution is used to model the number of defective items in samples drawn from large lots of items such as mechanical or electrical parts.
- Probability concepts are used in the stochastic design method for mechanical components (e.g., shafts, beams, plates and other solid objects under load). This is the design for reliability approach that is sometimes preferred over the design factor of safety approach.

- **Integration and Differentiation**

  Standard functions—polynomial, exponential, logarithmic (1st yr)
  Special functions—hyperbolic (1st)
  Product/quotient/chain rules (1st)
  Extreme points, points of inflection, and similar analysis of functions (1st)
  Setting up integrals (1st)
  Definite integrals representing area, energy, work, etc (1st)
  Moments of inertia & finding centroids (1st)
  Know how to get to the "answers" (1st)
  Multivariate calculus—derivatives/integrations—curl, divergence, gradient (2nd yr)
  Exact differentials $Mdx + Ndy = dz$ (2nd)

- **Linear Algebra**

  Vectors
  - Properties—magnitude, direction, linear, angular (1st yr)
  - Vector algebra and manipulation (dot product, cross product, possibly gradient) (1st yr)
  - Vectors as coordinate system (1st yr)
  - Linear independence (2nd yr)

  Matrices
  - How to cast equations into matrix form (1st yr)
  - Matrix manipulation (addition, multiplication, inversion, transposition) (1st yr)
  - Special matrices: symmetric, sparse, banded (2nd)
  - Basic dimensional analysis (2nd yr)
    - Singular matrices
    - Determinant
    - Rank

  Solution of system of homogeneous equations (1st or 2nd yr)
  - Eigenvectors and eigenvalues

  Vectors as a "special" matrix (2nd yr)
  Understanding homogeneous vs. non-homogeneous (non-forced vs. forced) (2nd yr)

- **Ordinary Differential Equations**
  1st, 2nd and higher order
    Constant coefficient (1st or 2nd order)
    Linear (1st or 2nd order)
    Boundary and initial value problems (1st or 2nd order)
    $dx/dt + ax = 0$ implies time constant (1st or 2nd order)
    d^2x/dt^2 interpreted as natural frequency and damping ratio (1st or 2nd order)
  Laplace transform
- **Partial Differential Equations**
  What are PDEs?
  How to convert PDEs to ODEs (2nd yr)
- **Numerical Methods** (not necessarily as a separate course)
  Symbolic manipulators (Matlab, etc.) (1st or 2nd yr)
  Discretization concepts (2nd yr)
  Finite differences (2nd yr)
  Numerical integration (2nd yr)
    Euler's Method
    Simpson's Rule
  Root finding (2nd yr)
  Runge-Kutta methods for ODEs (2nd yr)
  Regression and curve fitting (2nd yr)
- **Complex Numbers, Variables, and Functions**
  Arithmetic and algebra of complex numbers (1st yr)
  Representations of complex numbers: polar, Cartesian, exponential (1st yr)
  Laplace functions (2nd yr)
  Fourier series (2nd yr)

Third Year of Undergraduate Mathematics

- **Functions and Graphs**
  Complex plane
  Bode diagrams
- **Matrix analysis**
  Advanced matrix operations: Gauss, LUD (2nd or 3rd yr)
- **Ordinary differential equations**
  Nonlinear equations
  Periodic forcing functions
- **Partial differential equations**
  2nd order linear
  Simple systems
  Numerical methods

- **Statistics**

  Sampling (vs. total population) and estimation

  Factorial design and design of experiments

  Hypothesis testing (e.g., quality assurance)

  Examples

  - Statistical design of experiments (factorial design) is used to determine the main effects in experiments involving multiple variables, like the effects of fuel octane level, rpm, and torque on internal combustion engine efficiency. Multi-variable regression methods are also applied in experiments like this.
  - Control charts (e.g., X-bar and R-bar) are used extensively for continuous quality control decisions about manufacturing processes.

- **Numerical Methods**

  Boundary value problems—2nd order

  Finite element analysis

Health-Related Life Sciences

CRAFTY Curriculum Foundations Project
Virginia Commonwealth University, May 18–20, 2000

Thomas F. Huff and William J. Terrell, Report Editors
Reuben W. Farley and William E. Haver, Workshop Organizers

Summary

The conference participants reached agreement on the following aspects of a productive interaction between mathematics and the life sciences:

1. We affirm the value of the **fundamentals of mathematics** for the life sciences. A mastery of basic mathematical concepts is required, and the mathematical way of thinking is an essential part of the training of future life scientists.
2. We recognize a need for a **core mathematics curriculum** for students of the life sciences. Recommendations are made here for this mathematical core, although it is recognized that in practice there will be different local, institutional implementations of this core.
3. We recognize the necessity for **flexibility in client discipline curricula**. There is "not just one math student" coming to mathematics from the life sciences. Some students will want and need to take mathematical sciences courses beyond the recommended core, and the program for these students in client disciplines of the life sciences should allow for this diversity through customized individual programs.
4. We encourage **integration in teaching efforts** whenever possible. Progress will require communication and cooperation between mathematics and life sciences departments.

Our discussion of curricular issues should be an ongoing, dynamic process to improve our understanding of the details of these basic points of agreement. The interdisciplinary dialogue must continue in order to help drive this controlled evolution.

Narrative

Introduction and Background

We affirm that mathematics offers a language for precise communication, a way of thinking, and training in reading and writing technical material. As noted by a life sciences participant, students can "gain confidence in their ability to do things because of the mastery gained through their mathematical training."

Throughout these recommendations, the definition of **mastery** of a mathematical concept recognizes the importance of both conceptual understanding at the level of definition and understanding in terms of use/implementation/computation:

Mastery = (Conceptual understanding) + (Implementation/Computation)

Basic Concepts that should be mastered include:

- variables, parameters, functions, symbolic notation
- limits and their use especially in definitions of derivatives and integrals
- iterations
- probability, conditional probability
- approximation (as contrasted with exact value)
- logic and mathematical thinking, generalization, deductive reasoning

At the conceptual level, students should be able to explain these concepts in words.

Many participants put special emphasis on the use of models. Models are a way of organizing information for the purpose of gaining insight and providing intuition into systems that are too complex to understand any other way. There should be discussion about limitations of models ("what models cannot do")—these aspects are not generally covered in standard mathematics curricula. Some participants felt this view of modeling is not given sufficient status in biology.

The Core Curriculum should include:

(a) Basic concepts course(s): The previous paragraphs detail some specific content for this coursework. We believe it is appropriate to emphasize the use of models and the "meanings" of the mathematical concepts involved. Participants noted that mathematical modeling requires "a solid mathematical background in techniques and structures."

(b) Statistics—with an emphasis on the use of models.

(c) Use of computers and mathematics—we recognize a need for integrating mathematical concepts and relevant computer calculations for solving problems—true mastery should be the goal.

Although mathematics in biology uses discrete mathematics frequently, we affirm the importance of standard introductory "continuous" mathematics such as the standard content of introductory calculus courses.

Flexibility in Curriculum in the Life Sciences is needed. Conference participants from the life sciences repeatedly affirmed that the value of mathematics is not restricted to the specific content of any particular course or combination of courses.

The conference participants recommend to client disciplines that:

(a) There should be a long-term view of integration efforts that build productive interaction between mathematics and life sciences. Life sciences students should be permitted and indeed encouraged to take *more mathematical sciences courses in lieu of courses within the client departments.*

(b) Some students may want and need to take additional courses beyond the recommended core material. The "most likely optional list" as discussed by conference participants included coursework with coverage of
   - trigonometric functions (anything from basic through Fourier analysis)
   - matrices and linear algebra, including least squares problems
   - elementary differential equations or difference equations
   - non-parametric statistics
   - computer modeling and graphics
   - discrete mathematics (combinatorics, set theory, sorting)
   - advanced dynamics (nonlinear phenomena, time series analysis, equilibrium, stability, oscillations and limit cycles, bifurcations)

In the words of one participant:

> *The flexibility needed for the benefit of life sciences students is that the content of specific courses is not as important as the package of courses a life sciences student is encouraged to take.*

Another scientist commented on the needs of students with academic ambitions beyond undergraduate level:

> *The current areas of biological interest will multiply and branch into further areas, and mathematical issues will be crucial in forming links between these areas.*

Integration in Teaching Efforts. Integration can refer to specific courses (team teaching), or to an entire program structure. Some participants advocated "more mathematics concepts in biology courses, and more biology concepts in mathematics courses." Some mathematics participants invited client disciplines to "say what they want to see included in core course content."

At the undergraduate program level, *life sciences students might be allowed a program track that includes substantial mathematics course offerings. And mathematics departments might offer program tracks with substantial life sciences course offerings* for students with interests in those areas.

Teachers and researchers working across traditional academic boundaries need support and encouragement. Successful interdisciplinary work often involves overcoming the difficulties of language or academic culture differences.

Understanding and Content

The working definition of mastery is:

Mastery = (Conceptual understanding) + (Implementation/Computation)

This definition is motivated by the participants' acknowledgement that there has been a distancing between these aspects of fundamental understanding of mathematical concepts.

What conceptual mathematical principles must students master in the first two years?

We recommend that the entire Basic Concepts list be included here. We recognize that mathematics is cumulative in its development, and consequently, at the undergraduate level, it is essential that we work to maintain and build students' confidence in the value of mathematics. Students in the life sciences must obtain the necessary prerequisite mathematical skills in order to succeed in their chosen fields.

What mathematical problem solving skills must students master in the first two years?

Students should have a mastery of all aspects of linear functions in one variable.

All students should master graphing and visualization skills, including the use of log scales. Some participants observed that "chemistry students tend to think more visually than physics students, who tend to think in terms of formulae."

Students should master the use of computers to implement mathematical concepts for problem solving, and the use of statistics and models, through the use of a higher level interface as described below in the **Technology** section. Indeed, in the future many computations that are now typically done by hand or not done at all, will be performed analytically on computers.

What broad mathematical topics must students master in the first two years?

Here specific topics are distinguished from the broader list of basic concepts given previously.

Students should have a mastery of exponential functions (including base e exponentials), and logarithm functions, including properties of these functions required for their successful implementation in problem-solving.

What priorities exist between these topics?

The first priority is the basic core coverage as outlined above, with next priority to areas of specialization.

What is the desired balance between theoretical understanding and computational skill?

Our definition of mastery requires that both theoretical understanding and computational skill be considered essential elements of the overall understanding of concepts. The desired balance between the two aspects is problem/question dependent.

Examples where hand calculation is natural and desirable:

- Manipulating a linear equation by hand should be possible by every student.
- Finding slopes and axis intercepts from a given linear equation should be doable by hand.
- The estimation of the log of a number may require hand calculation using properties of exponential and log functions.
- Some participants mentioned "a familiarity with orders of magnitude" in connection with what should be doable by hand.

Some participants noted that undergraduate biology students in the first two years "may not necessarily know how to solve a differential equation, but they should understand qualitatively what the various terms in the equation mean." This emphasizes the basic conceptual understanding involved at some expense to the implementation aspect of understanding.

Students should realize (be taught) that conceptual understanding helps guide the search for **what** needs to be computed as well as **how** it may be computed, while technology/computation skills help guide **how** that calculation may be easily implemented.

What are the mathematical needs of different student populations and how can they be fulfilled?

We recommend that these different needs be addressed through the use of the "optional list" of coursework beyond the core curriculum outlined above.

Technology

How does technology affect what mathematics should be learned in the first two years?

Computer technology affects:

(a) How we do things
(b) Ease of implementation of mathematics
(c) Ability to visualize and conceptualize relevant concepts
(d) Ability to design relevant "experiments"

What mathematical technology skills should students master in the first two years?

Students should master a higher level interface, e.g.:

- spreadsheet
- symbolic/numerical computation packages (e.g., Mathematica, Maple, Matlab)
- statistical packages

The specific packages are less important than mastery of the concepts common to each class of package.

What different mathematical technology skills are required of different student populations?

We acknowledge that different technology skills may be required of different student populations.

Instructional Interconnections

What impact does mathematics education reform have on instruction in your discipline?

Biologists are generally unaware of mathematics education reform efforts. If the current recommendations of this conference are enacted throughout the curriculum, then positive change will be facilitated in life sciences curricula.

How should education reform in your discipline affect mathematics instruction?

All areas of biology have moved from observational/descriptive approaches to more quantitative approaches. These changes overall in biology have prompted the current recommendations of this conference.

We recommend that client departments require the core curriculum.

How can dialogue on educational issues between your discipline and mathematics best be maintained?

Participants agreed that dialogue on educational issues between disciplines should be maintained.

In the words of an undergraduate science major whose current expertise is in job placement of bachelor's degree students in the sciences:

> *These people need statistics, attention to detail, the ability to think logically, and the ability to master a variety of mathematical tools. Excellent verbal and written communication skills are also needed, since problem solving is usually done in groups of three or more people rather than alone. More extensive exposure to mathematics (beyond minimal core) is desirable.*

Effective dialogue will require institutional support (money, time, resources) for workshops and conferences, and convening of interdisciplinary committees (including participation from outside, e.g., industry stake holders).

Here are some examples of ways to further this dialogue:

- further communication with human resources (e.g., job placement) professionals
- cross-fertilization at scientific meetings
- co-authorship of texts and papers
- joint instruction in courses by mathematics/biology faculty, with more biology concepts and problems in mathematics courses
- national professional society interdisciplinary interactions to gain additional input, both anecdotal and quantitative
- interdisciplinary workshops, e.g., on technology, or conduct a local version of this MAA workshop
- sabbatical exchanges

WORKSHOP PARTICIPANTS

James E. Ames, IV, Computer Science, Virginia Commonwealth University
Mark Bloom, Biological Sciences Curriculum Study
Rich Carchman, Phillip Morris Incorporated
Jan Chlebowski, Biochemistry & Molecular Biophysics, Virginia Commonwealth University
Avis Cohen, Biology, University of Maryland
Lindon J. Eaves, Department of Human Genetics, Virginia Commonwealth University
Arthur S. Edison, Biochemistry and Molecular Biology, University of Florida
Warren Ewens, Biology, University of Pennsylvania
Charles P. Hatsell, Biodynamic Research Corporation
Ronald L. Hayes, Neuroscience, University of Florida Brain Institute
Thomas Huff, Vice Provost for Life Sciences, Virginia Commonwealth University
Lemont B. Kier, Medicinal Chemistry, Virginia Commonwealth University
Daijin Ko, Biostatistics, Virginia Commonwealth University
Larry S. Liebovitch, Center for Complex Systems and Brain Sciences, Florida Atlantic University
Frank Macrina, Vice President, Office of Research, Virginia Commonwealth University
Roland Moore, Division Chair, Mathematics and Science, J. Sargeant Reynolds Community College
Lawrence H. Muhlbaier, Duke University Medical Center
Stephanie Miller, On Assignment Lab Support
Leonard A. Smock, Chair, Biology, Virginia Commonwealth University
Halcyon Watkins, Biology, Prairie View University

Mathematics Participants

Thomas Banchoff, President, Mathematical Association of America, Brown University
William Barker, Mathematics, Bowdoin College
Dennis DeTurck, Mathematics, University of Pennsylvania
Reuben Farley, Mathematics, Virginia Commonwealth University
William Haver, Mathematics, Virginia Commonwealth University
Deborah Hughes-Hallet, Mathematics, University of Arizona
Daniel Kaplan, Mathematics, Macalester College
Phillip MacNeil, Mathematics, Norfolk State University
P. N. Raychowdhury, Mathematics, Virginia Commonwealth University
William J. Terrell, Mathematics, Virginia Commonwealth University
David C. Wilson, Mathematics, University of Florida
Lee Zia, Division of Undergraduate Education, National Science Foundation

ACKNOWLEDGEMENT

Grateful acknowledgement is made to Virginia Commonwealth University, the host institution for the Curriculum Foundations Workshop on the Health Related Life Sciences.

Interdisciplinary Core Mathematics

CRAFTY Curriculum Foundations Project
U.S. Military Academy, West Point, NY, November 4–7, 1999

Chris Arney and Don Small, Report Editors and Workshop Organizers

Summary

Forty-four engineers, mathematicians, and physical scientists met to examine the future role of undergraduate core mathematics (required courses in the first two years of instruction). The participants, divided into four groups, focused on the following areas:

1. Interdisciplinary Culture
2. Anticipated Advances in Technology
3. Goals and Content of Core Courses
4. Instructional Techniques

Abstracts of the discussions of the individual groups are presented in the Narrative. Then the many findings and recommendations made during the workshop are summarized under the headings: Consensus, Concerns, and Major Curriculum Initiative.

Appended to this report are excerpts from the keynote address, Urgency of Engineering Education Reform, given by Dr. William Wulf, President of the National Academy of Engineers, in which he noted that academia has not kept pace with changes in the professions and is failing to educate students to be technologically literate. With respect to mathematics, he encouraged a curriculum reform that spent less time on continuous, deterministic mathematics and more time on discrete and probabilistic mathematics.

Narrative

The state of our academic environment, in particular the interdisciplinary culture, is of great concern. Barriers between departments and lack of communication between faculty restrict the understanding and development of students. The workshop group recommended breaking down these barriers, establishing partnerships, and improving core mathematics programs to serve partner disciplines in the development of their students. The improvement of the curriculum through interdisciplinary cooperation is important as a first step, yet it still faces numerous roadblocks. We must prepare students for a diversity of careers in a rapidly changing environment. To do this, we need to develop broad reasoning and critical thinking skills that can only be accomplished through interdisciplinary cooperation.

Technology is a driving force in curriculum reform and is also a source of controversy. The ability of technology to provide visualization, numerical solutions and approximations, closed form symbolic solutions, and iterations has shifted curricula from focusing on mechanics and techniques to focusing on setting up problems and interpreting solutions. In short, technology has refocused curricula on the modeling process and away from the solution process. The resulting reduction of calculation skills is the major source of the controversy surrounding the use of technology. Another strong reason for incorporating technology in our core-course presentation is to prepare our students for the technological world into which they will graduate.

Content choices, balancing theory with computation, and the diversity of the students in first year courses lead to fundamental questions concerning the intellectual goals of a mathematics curriculum. Developing students to learn how to learn on their own has become accepted as central to the set of curriculum goals. Although not identical in meaning, the phrases *life long learner*, *learning how to think*, *mental discipline*, and *learning the mathematical thought process*, all seem to be perspectives on learning how to learn. There is no consensus on what to teach, and opinions ranged from maintaining the status quo to replacing calculus with a new program focusing on modeling and inquiry.

Some question why very little mathematics developed in the twentieth century is found in core courses. Others suggest that core content should be influenced by the needs of downstream courses, saying mathematics is basically a process, not a collection of topics. The primary concern for many is not content, but how to develop students to become competent, confident, and creative problem solvers.

In a suggested, yet controversial curriculum, modeling and applications would replace calculus as the umbrella course, yet the curriculum would still include rates of change, accumulation, transformation and approximation. All these concepts and many more would arise from modeling realistic situations rather than from studying specified topics. An important aspect of this projected modeling program is the integration of data analysis, probability, and discrete mathematics with continuous mathematics, since the situations being modeled rarely fall into our artificial curriculum categories. Another important aspect is that a modeling program is inherently interdisciplinary because real-world situations are interdisciplinary.

Student growth should be accounted for in curriculum planning; it is too important to be left to chance. Identified areas for attention are learning how to learn, communication, mathematical sophistication, modeling, technology, connection with other disciplines, and history of mathematics. The meaning of high standards in core courses at West Point has changed from preparation for upper-level mathematics concepts in real analysis to new standards that relate to deeper modeling experiences, open-ended projects, inquiry, and the ability to apply mathematics in interdisciplinary settings.

The workshop participants applauded the shift of focus in courses from teacher-centered to ones being learner-centered with less topical coverage and greater depth in the content. The pedagogical shift involves engaging students in multiple learning activities such as group activities, group projects, discovery work, technology laboratory sessions, writing assignments, and student presentations. The difficulties in performing this pedagogical reform lie in the fact that many student-learning activities take considerable class-time, sometimes require more work for the instructor, and often require a change in assessment methods. The time factor poses the biggest challenge. The undergraduate teaching profession has been reluctant to reduce content in order to make time for student learning activities and has difficulty accepting that learning is a very inefficient process. Somehow, this reluctance must be overcome and the center of gravity of the core program must continue to move.

Nine instructional methods/issues were analyzed and recommended as valuable instructional techniques in core mathematics programs: Questioning/Discussion, Problem Solving, Use of Technology, Exploration and Discovery, Multiple Representation, Writing, Multiple Assessment Instruments, Control of Section Size, and Use of Group work.

Consensus

Workshop participants achieved consensus on the following desired attributes of an undergraduate mathematics curriculum, especially when significant interdisciplinary connections are involved.

1. More *modeling* should be incorporated into the curriculum.

   Modeling was viewed by each of the subgroups as an effective means of addressing their issues. The Interdisciplinary group saw modeling as representing the best approach to breaking down current barriers to interdisciplinary cooperation. Technology is seen as moving curricula toward the modeling process and away from the solution process. Although in agreement on increasing the emphasis on modeling to prepare students to become competent, confident, and creative problem solvers, the members of the Goals and Content group differed on the extent. The Instruction group viewed modeling as an effective way to address their multiple learning issues.

2. There should be a greater emphasis on *problem solving* in the sense of modeling real-world problems rather than in the sense of exercises.
3. The curriculum needs to emphasize *learning how to learn.*
   Although not identical in meaning the frequently heard phrases: *life long learner*, *learning to think*, *mental discipline*, and *learning the mathematical thought process* all offered perspectives on learning how to learn.
4. Instructors must make *effective use of technology*, particularly for visualization, discovery, and insight as well as computation. Students must be prepared for the technological world into which they will graduate. The terms *effective use* and *appropriate technology* need to be better defined.
5. There should be a pedagogical shift from teacher-centered instruction to *learner-centered instruction*. Topical coverage can be reduced, with remaining material developed to greater depth.
6. There is great value in the use of *multiple learning activities:* projects, discovery work, writing, presentations, calculator/computer laboratory sessions.
7. *Process* is more important than *content*.

Concerns

Workshop participants identified the following problems and issues which need to be confronted when revising an undergraduate mathematics curriculum, especially if significant interdisciplinary connections are desired.

1. The present state of interdisciplinary cooperation and interaction needs improvement.
   Barriers between departments and lack of communication between faculty restrict student development. Although there is (theoretical) agreement on the benefits of interdisciplinary cooperation, several barriers exist such as system inertia, fiefs and turfs, publish or perish syndromes focused on narrow results, entrenched attitudes, rigid reward systems, and time. The low intensity of interdisciplinary cooperation has restricted reform efforts in mathematics, physics, and engineering.
2. There can be a loss of calculation skills due to the use of technology.
3. It is not always clear how to use technology effectively.
4. Serious time issues are raised by the recommended curricular and pedagogical revisions.
   Student-learning activities take time away from instructor activities and require more instructor time (developing materials, grading complex projects) and different means of assessment. There can be a major conflict between content coverage and student-learning activities: such activities can be time intensive, reducing the number of topics that can be covered in a course.
5. There is a lack of agreement on content choices and priorities (continuous vs discrete, deterministic vs stochastic, etc.)
6. There is often a conflict between the *math way* (e.g., emphasizing limits as the major *primitive*) and the *science way* (e.g., emphasizing rates of change as the major *primitive*).
7. There is an unfortunate lack of statistics/data analysis in the introductory courses.

Major Curriculum Initiative

Mathematics departments should consider adopting a core sequence of courses focused on inquiry and modeling that interweaves continuous and discrete mathematics. Calculus topics of rates of change, accumulation, transformations, approximations, and others would arise through modeling realistic situations rather than studying specified subjects. Similarly data analysis, statistics, probability, graph theory, matrix algebra and other discrete topics would also arise through modeling realistic situations. The program would more effectively address the goal of developing competent, confident, and creative problem solvers than do the present calculus courses. In addition, the program would be inherently interdisciplinary, as real-world situations are interdisciplinary.

REFERENCES

The entire Proceedings of the workshop, including the 35 submitted position papers, can be found at `www.dean.usma.edu/math/activities/ilap/workshops/1999/default.html`

Additional information and details compiled from the workshop are found in *Changing Core Mathematics*, MAA Notes Volume #61 (Edited by Chris Arney and Don Small). This volume contains a historical description of the evolution of the mathematics curriculum and an expanded description of the major curriculum initiative presented at the workshop.

WORKSHOP PARTICIPANTS

Don Albers, Mathematical Association of America

Chris Arney, Professor of Mathematics, U.S. Military Academy

William Barker, Professor of Mathematics, Bowdoin College

Lida Barrett, Professor of Mathematics—retired, U.S. Military Academy

Tom Berger, Professor of Mathematics, Colby College

Ray Cannon, Professor of Mathematics, Baylor University

Jim Daley, Professor of Engineering—retired, University of Maryland

Lisette dePillis, Professor of Mathematics, Harvey Mudd College

John Dossey, Professor of Mathematics, Illinois State University

Pat Driscoll, Professor of Mathematics, U.S. Military Academy

Penny Dunham, Professor of Mathematics, Muhlenberg College

Mary Goodwin, Professor of Engineering, Iowa State University

Laurette Foster, Professor of Mathematics, Prairie View A & M University

Jeff Froyd, Professor of Engineering, Rose-Hulman Institute of Technology

Bob Fuller, Professor of Physics, University of Nebraska

Frank Giordano, Professor of Mathematics, COMAP, Inc.

Jack Grubbs, Professor of Environmental Engineering, Tulane University

Gary Krahn, Professor of Mathematics, U.S. Military Academy

Heidi Mauk, Professor of Physics, U.S, Air Force Academy

Bill Haver, Professor of Mathematics, Virginia Commonwealth University

Tom Lanis, Professor of Physics, U.S. Military Academy
Jim Lightbourne, National Science Foundation
Dave Lomen, Professor of Mathematics, University of Arizona
Mike McGinnis, Professor of Engineering, U.S. Military Academy
Joe Myers, Professor of Mathematics, U.S. Military Academy
Charlie Patton, Science, Mathematics, Technology Service
Richard Plumb, Professor of Engineering, State University of New York–Binghamton
Shirley Pomeranz, Professor of Mathematics, University of Tulsa
Fred Rickey, Professor of Mathematics, U.S. Military Academy
Andre Sayles, Professor of Engineering, U.S. Military Academy
John Scharf, Professor of Engineering, Carroll College
Bob Soutas-Little, Professor of Engineering, Michigan State University
Don Small, Professor of Mathematics, U.S. Military Academy
Kathi Snook, Professor of Mathematics, U.S. Military Academy
Jim Stith, Professor of Physics, American Institute of Physics
Liz Teles, National Science Foundation
Bill Vanbuskirk, Professor of Engineering, New Jersey Institute of Technology
Frank Wattenberg, Professor of Mathematics, U.S. Military Academy
Bill Wilhelm, Professor of Mathematics, U.S. Military Academy
Brian Winkel, Professor of Mathematics, U.S. Military Academy
Bill Wulf, Professor of Engineering, National Academy of Engineering
Lee Zia, National Science Foundation
Paul Zorn, Professor of Mathematics, St. Olaf College

ACKNOWLEDGEMENTS

The U.S. Military Academy hosted the Interdisciplinary Workshop on Core Mathematics. We express our appreciation to the Department of Mathematical Sciences and several administrative assistants: Nelson Emmons, Connie Arndt, Loren Eggen, and Andrew Weate.

The National Science Foundation supported the Interdisciplinary Workshop on Core Mathematics through Project INTERMATH, as part of the Mathematics Across the Curriculum initiative.

APPENDIX: The Urgency of Engineering Education Reform

Exerpts from the Keynote Address by William Wulf, President of the National Academy of Engineering

I want to talk about engineering education and what I sense is the real urgency of engineering education reform. I think we ought to be seeing a watershed change in engineering education—it is not happening. I am very impatient about it and I hope I can communicate to you why I feel impatient about it. A lot has been written on the subject. There were a whole series of reports done in the 1994–1995 timeframe. There was one done by NSF, there was one done by the National Research Council; and there was one done by the Dean's Council of ASEE (American Society for Engineering Education). All of them called for a fairly dramatic reform.

I have three introductory remarks to make before engaging in talking specifically about what I think needs to be done. First, a caveat, I am going to paint with a very broad brush. I fully appreciate that if you go to any engineering school you are likely to find some innovative things happening. What is not happening is the center of gravity moving in any substantive way. That is my concern.

Second, I have a particular view of what an engineer does that colors the way that I think about these things. I want to contrast it with science for a moment. Science is fundamentally analytic. Its concern is with the understanding of nature—understanding what "is". Engineering is fundamentally synthetic. It is concerned with creating what "can be". That difference in approach is profound. My favorite operational definition of what an engineer does is "design under constraint". Given a problem an engineer designs a solution, but not any old solution will do. You have to satisfy a set of constraints—and I will argue in a minute that that set of constraints is getting much more complicated. You have to worry about, first of all, functionality—solving the problem—but then you've got size, cost, weight, heat dissipation, and on and on—I will talk about this more later. If you really want to get my ire up, say that engineering is just applied science. Engineering is not just applied science. Engineering is philosophically at its core very different. It is fundamentally creative rather than explanatory. To be sure, our understanding of nature is one of the constraints that an engineer works under. In my personal experience in the company I founded and ran, it turns out that nature is almost never the limiting constraint. Our understanding of nature is seldom the hardest constraint that you work with.

The third caveat, and maybe this is the most important one—engineering is changing. Indeed it's that change that underlies my sense of urgency in the need for engineering education to change. I believe that the way that we will practice engineering and the way that the students we are teaching today will practice engineering are profoundly different from the way that I practiced engineering or my father practiced engineering. The problem with trying to describe to you what that change is about is rather like standing too close to a mosaic. I have said, sometimes there are monumental events that kind of cast a sharp knife edge between the way things were and the way things are now. World War II strikes me that way. Before World War II there was no federal funding of research at universities. After World War II we built this wonderful mechanism for funding research. The role of women in society dramatically changes across that boundary. In fact, engineering education changes dramatically across that boundary. The notion of the engineering-science model of engineering education comes about because of, frankly, the failure of engineers to contribute as much as scientists did to the war effort.

I don't think we are in that kind of a change. I don't see that monumental event. It seems to me that this is much more like the Industrial Revolution. You know, we talk about the Industrial Revolution now as though it was an event. The fact is, it smeared out over almost 100 years and it is contemporaneous with a whole bunch of profound changes in society. This is when you get the rise of democracy; this is the rise of rationalism; and there was another great change in university education. The introduction of liberal or secular education comes about at exactly the same time. If you were there at the time, you could not have predicted what the world would look like at the end of that time. I think we are in that kind of change.

So I am going to be describing bits of this mosaic to you as opposed to "I'll tell you what engineering practice is going to be 20 years from now"—I haven't the foggiest idea. I can just tell you there are these forces that are, I think, dramatically changing things. I see at least six pieces to this mosaic of change that I want to talk about today. First, I said engineering is designing under constraint. So I want to talk about: The complexity of the design space which I think is exploding. The complexity of the constraint set which is also exploding. I want to talk about what I will call "the fallacy of the possibility of precision". Then I want to talk about a couple of social changes in engineering. The expanding role of engineers in industry, the globalization of engineering, and then note that the pace of change is in itself a change.

Let me talk about the complexity of the design space. When I say design space, what I mean is, for each decision that an engineer makes: How big will this thing be? How heavy will it be? How much power consumption can I allow this thing to have? For each such decision, you want to think of that as a dimension in a design space and each option that the engineer has as a point along that dimension. So each point in that space is a potential solution to the problem that you are trying to solve. It may be a good solution, or it may not be a good solution, but it is a potential solution.

Let me just illustrate with three examples why I say the design space is getting much more complicated. The examples are: materials, information technology, and systems—and I am not going to say here anything that you don't already know. My father was an engineer. He was a mechanical engineer. He designed machines for a company that made cookies. I can remember growing up and going to his plant and just being amazed at how you could get very flaky crackers, for example, to be mass-produced at a horrendous pace. I mean they just came flying out of this literally 300 ft long oven. But, for my father there was a little book on a shelf, a little thin book, of the materials that he had as an option to design with. There were a half a dozen different kinds of steel, there were a few kinds of bronze, plastic was not in his vocabulary, fibers were not in his vocabulary, composite materials were not something he considered.

Well, now we are talking about designer materials, which give an engineer the ability to say "these are the properties that I want the material to have" and at least potentially the possibility of producing that material for that subject. Literally, that thin book has become an infinite set of options. The notion of biomaterials (you know we talk about biotechnology a lot in terms of medical applications), but do you know what the slipperiest stuff in the world is? The stuff with the lowest coefficient of friction known to man? It is the stuff at the end of your bones. There is no man-made material as slippery as your joints. We are going to be talking about growing materials. One of my colleagues at Virginia is into making smart materials and it is almost scary. He talks about materials that understand their role in a structure, sense the environment, and adapt their properties to better fulfill their role. Materials in which electrical properties and small forces can be exploited to build structures that are very much lighter and do in fact adapt to their environment with very small changes.

My second example is Information Technology: Everybody knows Moore's Law? Two times the number of transistors on a per unit area every 18 months. The fact is you can have intelligence imbedded in everything. There will not be a product produced 20 years from now that doesn't have some degree of intelligence. Have you ever played this game of how many electric motors you have in your house? You know, as we went through the transition from watermills to steam engines, in both cases plants had these great big shafts down the middle of the plant and hung belts off of them to run all the machinery. The first use of electric motors was simply to replace steam engines that ran the shaft. Then slowly every tool got its own electric motor, and now, of course, we just embed an electric motor in everything. The typical home has hundreds or thousands of electric motors. I was standing in the shower one day wondering how many computers I had in my bathroom. I know of at least two and I probably don't know of some others. Because its the cheap way to imbed control into a product.

One of the projects I was working on in Virginia before I took on this job dealt with bridge construction. Building a bridge is expensive. Inspecting a bridge is even more expensive. The rebar and the concrete slowly corrode. Concrete cracks and water seeps in onto the rebar. So you have to inspect the bridge to make sure the concrete is still doing its thing. We were designing a chip that contained a corrosion sensor, a microprocessor, and a small radio transceiver. Objective—make it cheap enough that you can put a shovel-full in every load of concrete and simply drive a truck across the bridge with a radio transmitter that asks the bridge whether it is corroded or not.

Everything is going to have intelligence imbedded in it—everything. If you start thinking of combining IT with MIMS, the potential is absolutely incredible and I haven't even started talking about nano-technology yet.

The third thing I want to mention with respect to complexity is systems. Simply; the number of components per product has been going up exponentially and we are starting to hit that point of the curve where it really, really is going to go up fast. That is going to imply more and more kinds of engineering expertise to produce any single product. So, the bottom line is that the design space, the number of options that an engineer has, is just going through the roof.

Design under constraint—the design space is going up—I want to argue that the complexity of the constraint set is going up equally and rapidly. My father had primarily two constraints to work under—functionality and cost—one of those was a fixed point—the machine had to work, so he was designing against one free variable. This is particularly true when you are building great big machines. It doesn't matter whether the thing weighs 200 lbs or 400 lbs except to the extent that weight represents additional cost. Well, if you look at our society now, the constraint set includes safety, reliability, manufacturability or remanufacturability, repairability, maintainability, a whole set of ecological concerns that didn't exist before, ergonomic concerns that didn't exist before, human interface considerations that we never thought about before, and many, many more things.

It is not only that the list of constraints is huge: the optimization function isn't clear. For my Dad, fixed point functionality—drive the cost as low as you can—easy optimization function. Not at all clear what the optimization function is for things like ecological concerns. We have time after time found that driving down the knocks in automobile emissions does not necessarily minimize pollution in places like the Los Angeles basin. It is a much more complicated chemical process. Not only that, but you don't even know how to measure some of these things. What are the units of ergonomics suitability? Oh, and by the way, the public seems to believe that some things are absolutes. No degree of environmental impact is acceptable. There is no lower bound on what the public is willing to accept.

So the argument I am trying to make to you is the design space has gotten much bigger, the constraint set has gotten much bigger, and it's a different kind of engineering world than it was for my Dad. Not only that, it is not even clear what constitutes the best design.

Now let me talk about the possibility of precision. For my Dad, looking back in particular, I realize he was a very good engineer. But there was absolutely no way that he could *a priori* predict what the exact behavior of his machine would be. I mean, it was just a given that you built a prototype and it might work as intended, but probably it wouldn't. You would probably wind up having to modify some things in it. That was the whole idea of building prototypes. You worked with kind of crude orders of magnitude computation. He had a lot of knowledge of prior machines that he drew on, but basically nobody expected the first thing out of the pocket to work, and in fact if the first one didn't work his boss wasn't upset about that. There wasn't someone standing in line to sue him because of it.

But with modern computation, and better and better models of the physical world, a better and better understanding of the physical world, it is in fact apparently possible to be precise. Everybody talks about the Boeing 777, for example, for which no prototype was built. The first one that was built was the first one that flew and that was because of the modeling that was done ahead of time. At least in principle it appears possible to be precise.

Now I claim that this is kind of a mixed blessing. On the one hand it is nice not to have to build a prototype, but it carries with it an implied responsibility. It is not unreasonable for your boss, your insurer, your customer, the federal regulator, to believe that the first prototype will work as intended. Now what does that mean? That means that in the face of this much more complicated design space, much more complicated constraint set, you as an engineer have an implied responsibility to search all of it, to make sure that the design you come up with is really the global optimum in that space.

Well, I frankly just don't think that is possible. I happen to be a computer guy, as I was introduced. Can I teach a little bit of computer science for a minute? I am going to wave my hands so if I bore you forgive me. You have all seen programming languages. You all know they contain classes of statements. For each one of those classes of statements it is possible *a priori* to specify the following. Suppose you had a logical expression that characterized the state of the system after the execution of the statement. It is possible to mechanically take that logical expression and the statement and produce another logical expression that must have been true in order for the statement to have been executed and to result in the logical expression that follows. If you have an assignment $x = y + z$, then for any property that was true of x after the statement was executed, that same statement must have been true for the expression $y + z$ before the statement was executed. I can do that for every kind of statement in the programming language. What that means is if you can write down a logical expression that describes the state that you want to be true at the end of the execution of a program, I can completely, mechanically, and really quite simply back that statement up, that logical expression, one statement at a time through the program and derive an expression which must be true at the beginning of the execution in order to get the right thing at the end. Well, if that expression at the beginning is a tautology, if it's always true, then I can absolutely guarantee that the program works right. Possibility of precision—it is possible to write programs which absolutely are guaranteed to be correct. Or at least produce the results that you said you wanted.

And yet have you ever encountered a program that was correct? I rather suspect that you haven't. Now, why is that? Well, there are two things—first of all the logical expression that you get at the beginning after doing this backing up is huge and our ability to prove those to be tautologies is not up to the task. But there is something more important than that. Humanly, we are not able to describe what it means to be correct. We are not able to write the expression that you want at the end.

One of my research areas—I've had a crazy research career doing lots of different things—one of them has been computer security. Within the domain of computer security there are things called cryptographic protocols. Cryptographic protocols happen to use cryptographic techniques, but they are intended for situations like—if you have two parties at opposite ends of the communication lines, each party should be able to verify that the party they are talking to at the other end is who they claim they are. I want to be sure that I am talking to you, and you want to be sure you are talking to me. These protocols are often ten-line programs. They are really tiny. They are something you think you can verify. And in fact people have published proofs of the correctness of a number of these protocols which have subsequently been shown to not work. In essence because it is just much harder to describe what constitutes correctness than you might think. So, I find this possibility of precision to be one of the things that may have the most profound effect on the practice of engineering. We are going to be expected to be precise in an environment where it is not at all clear whether that is an achievable goal. .

I would like now to talk about the expanded role of engineers in industry. Everybody has written about or heard and read about teams. About how industrial practice now is very much oriented around marketing people, engineers, financial people, etc., working together on a product. That is an environment in which the engineer we are training today is not equipped to operate. When I first heard about it, I thought it was a passing fad. The more I think about it, the more I realize that that's the way engineers operated through almost all of history. The era of specialization, of having an engineering department that threw designs over a transom, is the anomaly. Now whether the particular management fad of the day on how you do that will persist—no I don't think so. But the notion I think is enduring. Globalization of industry maybe is a special case of a team thing. Lots of people are more expert at this than I am, but it seems to me that this really underscores the fact that the engineer who is trained superbly in a technical sense, but does not understand the cultural and social issues in a very broad sense, in a multicultural way, is really useless.

Another important perspective is the pace of change is itself a change. Just as I came on board for this job, the NAE was concluding a conference about life long learning in engineering and somebody at that conference talked about the half-life of engineering knowledge. How long does it take for half of what an engineer knows to become obsolete? I must admit I quote these numbers all the time without ever verifying them, just because the dramatic effect is worth it. I won't stand behind these numbers, but what was estimated at that conference was that it varies by field from 7.5 down to 2.5 years. It so happens in software engineering:that the claim was half of what you know becomes obsolete in 2.5 years. Frankly, I am a little uncomfortable with that kind of one-dimensional characterization, but the important point is that it has not been part of the engineers culture to feel responsible for their own life long learning and I think that has to change.

There is a bunch of stuff that needs to change: curriculum, pedagogy, (I am particularly sensitive to the issue of diversity), the notion that the baccalaureate is the first professional degree, faculty reward system, the need for formalized lifelong learning, preparation in K through 12 and technological literacy in the general population

Let's talk about the first professional degree. Whether you are talking about medicine, law, business, there is no other profession that treats the baccalaureate as the first professional degree. And I think, frankly, the fact that we do causes all kinds of foolishness. We misrepresent the situation to both the students and potential employers. We seem to be perfectly comfortable with the notion that an employer is going to spend the first couple of years adding to the education of our products before they are useful. It has caused our curriculum to expand to the order of 135 semester hours as compared with 120. And by the way, that problem is going to become truly acute when states like my own, Virginia, actually do what they say they are going to do, namely mandate that the engineering program be a 120-hour program. We are going to lose five courses out of the curriculum. We'll squeeze out the humanities, liberal arts, which I think are becoming central to what an engineer is going to have to be able to do.

You may not know this, but engineering is not a profession. We may like to talk about it being a profession, but in a technical sense the Department of Education defines what is a profession and there are two properties that a profession must have. The first one is at least two years post-baccalaureate. Second, it has to be on "the list". The DOE maintains a list of the professions, and engineering is not on that list. My members are quite offended that they are not considered professionals, but technically they are not.

Curriculum—if you get a bunch of engineers together there is an oath that we all recite. That oath is that what we must do in the baccalaureate is teach "only the fundamentals". "Only the fundamentals"—you hear that recited over and over again because we treat the baccalaureate as the first professional degree. Well, rubber meets the road when you ask what are fundamentals? And then the mechanicals will tell you something quite different from the civils, and neither one of them will recognize, for example, that they sort of agree, because since WWII the fundamentals have included continuous mathematics and physics. That much I think everybody agrees on.

But as I said before engineering is changing. Information technology is going to be imbedded in everything that engineers produce. And discrete mathematics, not continuous mathematics, is the underpinning of information technology. I mentioned biological materials. Biology and chemistry are going to become as fundamental as continuous mathematics and physics. And the fact that engineering is done in this more holistic team-oriented, multinational global context means that there are a whole set of business and cultural issues that are really fundamental to engineering. You can't practice without them.

If you want to continue to say that the baccalaureate is the first professional degree, then you have to agree that some of our cherished current fundamentals aren't any more. Or you have to figure out a way to teach them much more efficiently and effectively. I happen to think that continuous mathematics ought to be done in two semesters, not four, and I think that is possible to do. But I leave that to all of you to figure out how to do.

While I am on the subject of curriculum let me come back to the possibility of precision for just a minute. One of the properties that we see in software systems, and I think you will see in all engineering systems as they become increasingly complicated, are what are called immerging properties. The systems behave as specified but they also have other properties, other behaviors that you did not anticipate. The question is how do you engineer safe, reliable, cost-effective products whose behavior you could not have anticipated ahead of time. It is not that you are a lousy engineer, that you did a bad job, it's that you literally could not have anticipated everything ahead of time. The complexity of the system is such that it is infeasible. I think this is an opportunity for a whole new class of mathematics. Don't ask me what it would be.

Ethics has been very important to engineering. Engineers are very much like physicians—first do no harm. We spend a lot of time teaching engineers how to over-design their systems so that they tend to not fail or if they fail, fail safe. How dc you cope with the ethics of not knowing what the behavior, what the immerging properties, of a system will be? I don't know.

Let's talk about faculty rewards. And I don't mean the teaching versus research debate. I happen to be one of those people who believes that, most of the time, research and teaching compliment each other. Most of the people who I know who are good researchers are also good teachers. Good people are good. There are the outliers. But I think we have another problem. Remember I said I believe what engineers do is design under constraint. I happen to think that engineering is an incredibly creative activity. Something we don't advertise very well. In my heart, I believe that engineering is one of the most creative of human activities. If you stipulate that for just a minute, can you think of any other creative activity, on campus, where you don't expect the faculty to practice, to perform that creative activity? The Art Department doesn't promote or tenure anybody who doesn't practice their art. Think about the Music Department. Or even think about the other professions like law and medicine. If you go to medical school, you go on grand rounds with the faculty who is practicing his/her profession. Engineering is the only creative activity that I can think of where, in fact, the faculty are actively discouraged from practicing the profession. And what we wound up with—you know the criteria that we apply for promotion tenure in universities is essentially derived from the Science Departments. The criteria are research, publication, getting grants—and you'd better teach pretty well too. But, practicing the profession counts for nothing and probably counts against you because it detracts from other things.

I actually had a Dean who would not let one of my faculty take a sabbatical in industry. His belief was that there was nothing to be learned from industrial experience, and in fact somehow those industrial people were just going to suck out his brains and take out everything he knew. Well, I can tell you, I spent almost ten years of my life in the private sector and one of the most intellectually challenging things I have ever done in my life was delivering product. It is not just that it is hard, it's intellectually challenging. Going back to the curriculum issue for just a moment, I think one of the things that is really wrong is that we have a curriculum being designed by faculty members who are not practicing engineers. I have a great deal of respect for my colleagues at the university. They are wonderful engineering scientists, but very few of them know anything about the practice of engineering,and so they design a curriculum that is an engineering-science curriculum, not an engineering-practice curriculum.

Let me talk about the notion of technological literacy in the general public. Before I took this job, I was a Professor at the University of Virginia. As many of you may know, Virginia was founded by Thomas Jefferson. What you probably don't know is that Jefferson did not die, he participates actively every day in the decision mechanisms of the university. He was very proud of having founded the university. It was one of three things he put on his tombstone. He didn't mention things like being President of the United States. He founded the university because he believed you could not have a democracy without having an educated citizenry.

Well, I think he would be scared today because we have a citizenry that is not only ignorant of technology, it is proud of the fact that it is ignorant of technology. You know, I go to a cocktail party and someone will ask me what I do and I say I teach computer science and they say, "Oh, I don't understand that computer stuff". Can you imagine asking somebody else what they did and they said they were a Professor of English and you say "Oh, nouns and verbs, I can't …." Engineering schools don't offer technological literacy courses for liberal arts majors. Why not? We could pass on knowledge of not just science and math but the process that takes that knowledge of nature and converts it into the things that profoundly change our quality of life. Think about how the average person in 1899 lived. Think about how an average person in 1999 lives. All of the differences are engineer products. In 1899 the average life span was 46. In 1999 the average life span is 76. Not all of that increase is due to modern medicine. It is almost all due to cleaner water and sanitary sewers—public health. Engineering!!

And yet, "Oh, I don't understand that computer stuff and I am proud of the fact that I don't." Every person who has a liberal education ought to be at some level technologically literate and it's our responsibility to provide the opportunity for that to happen. It is no good to point a finger and say "You English professors ought to be technologically literate" if there is no mechanism for them to do that.

I've tried to indicate to you that I think the practice of engineering is going to change tremendously and that therefore the education of engineers needs to change tremendously. I love this quote: Wayne Gretzky, probably the best hockey player that ever lived, talked about the fact that he didn't skate to where the puck was, he skated to where the puck would be. I'm afraid that engineering education is skating to where the puck was.

Mathematics

CRAFTY Curriculum Foundations Project
Mathematical Sciences Research Institute, February 9–11, 2001

Herbert Kasube and William McCallum, Report Editors
William McCallum, Workshop Organizer

Summary

The most important task of the first two years is to move students from a procedural/computational understanding of mathematics to a broad understanding encompassing logical reasoning, generalization, abstraction, and formal proof. This should be achieved as soon as possible in a student's undergraduate career.

The transition should be accompanied by an increased sophistication of student attitudes towards mathematics. For example, students should develop an understanding that they can learn from failure. They should come to understand that solutions are often multi-staged and require significant creativity, time and patience, and that they can gain considerable insight by analyzing whatever went wrong. Courses should instill confidence that mathematics makes sense, is worthwhile, and has intellectual vitality.

Narrative

Essential themes—fundamental ideas or principles that apply to varied subject matter—should be threaded through different courses. Departments should think about how to provide coherent integration of these themes across courses. Some themes are about mathematics as a whole:

- The nature of mathematical language: basic logic, the difference between a statement and its converse, semantics of quantifiers, the role of definitions.
- The nature of mathematical knowledge: the nature of proof, why theorems are important, why the universality of a statement is important in mathematics.
- Interrelations between concepts, skills, and topics.

Some themes are more specific:

- Concept of function (as distinct from formula).
- Approximation.
- Algorithm.
- Linearity.
- Dimension.

Students should gain proficiency in the following skills in the first two years:

- Understanding and doing standard computations
- Visualization and geometric skills, especially 3-dimensional visualization and solid geometry.
- Translating mathematics into words and translating words (statement of problems) into mathematics.

- Generalization and checking general statements via specific examples and experimentation.
- Recognizing invalid arguments and incorrect answers; knowing when the result/answer is reasonable.
- Analyzing the problem (is it well defined?), attacking the problem (starting without knowing the answer), using all techniques and results available (not quitting after one attempt).
- Stating problems clearly and setting them up (e.g., defining variables, deciding which variable is to be solved for, setting up coordinates, defining notation).
- Communicating mathematics in writing and orally, using precise reasoning and genuine analysis.

There should be a balance between the development of computational skill, conceptual understanding, theoretical reasoning, and applications. Connections between these should be brought out: computational technique can be used to illustrate and develop theory, and applications can be used to give flesh to conceptual understanding. There has been, on average, an overemphasis on developing computational skill in the introductory courses. Similarly, in advanced courses there is often a stress on theoretical reasoning before the necessary conceptual understanding has been achieved. There should be an attempt to phase in logical language starting in the freshman year, rather than a sudden jump in the sophomore or junior year.

Calculus and linear algebra are likely to remain the primary core of the first two years of undergraduate mathematics. We should strive to make such courses reflect the core of mathematical culture: the value and validity of careful reasoning, of precise definition, and close argument. Doing so in an appropriate and effective way is the challenge of the current decade.

Students should have the opportunity to sample a wider variety of topics. Such topics should include mathematics that is of recent origin or current interest, illustrating our discipline as a continually growing and developing body of knowledge, driven by creativity, a passionate search for fundamental structure and interrelationships, and a methodology that is both powerful and intellectually compelling. Students should encounter mathematics in an interesting historical, aesthetic or useful context. Prospective high school teachers need a deepening of their understanding of high school mathematics. Alternatives to calculus and linear algebra should be available in the first two years. Some possibilities are: discrete mathematics, number theory, geometry, and knot theory.

Realism dictates recognition that students are headed in many directions besides graduate school in mathematics: K–12 teaching, graduate school in various other disciplines, industry, business, and government. Many do not yet have clear plans but have a general interest in mathematics. It is unrealistic to suppose that students can always be sorted according to their different goals, but we should prepare students to start making decisions about different tracks that might open up in the last two years.

Similarly, we should recognize our students' great variety of individual, cultural, and educational backgrounds. Students come with vastly different experience, skills, and learning styles. Introductory courses should provide experiences flexible enough to allow every student the opportunity both to reinforce existing strengths and to fill gaps.

Technology

Prospective mathematics majors should be familiar with appropriate use of technology. They should, in particular, have experience with a computer algebra system during the first two years of undergraduate training. In addition to its obvious value for visualization and for lengthy computations, technology is useful as a diagnostic tool. Instructors can also use technology as a basis for student experiences requiring active thought. In addition, there are creative web applications to support mathematics learning such as:

- JOMA, the Journal of Online Mathematics and its Applications
- ALEKS, the Aleks Corporation

The existence and ubiquity of high-powered computational technology reduces the importance of student mastery of manual computations of an intricate and specialized nature. Since the interactions between computational fluency and broader learning goals are not understood, caution is needed to avoid loss of technique that would impede later mastery of mathematics. This probably depends to some extent on the future plans of the student. However, there is a concern that introduction of technology is often accompanied by a decrease in computational fluency.

It should be observed that graphing calculators have limited uses outside of the mathematics classroom. Though there are exceptions, graphing calculators are little used, professionally or pedagogically, in other disciplines. A decision to use graphing calculators cannot therefore be based solely on unsubstantiated reports of their ubiquity or general usefulness in other domains.

Most instructors are too busy to investigate new technology. However, it is in a department's best interest to acquaint itself with promising new developments. A mathematics department should take some responsibility for helping its instructors understand the different platforms for the teaching of mathematics.

Instructional Formats and Techniques

For the purpose of this report, we distinguish between instructional *formats* (such as large lecture, small group work, laboratories) and instructional *techniques* (such as quizzes, student presentations, written homework).

Instruction should be viewed as interaction rather than a one-way process. Instructional techniques should be employed that foster discussion, improve the value added by class time, encourage student interaction (either inside or outside the classroom), engage the students in the subject and ignite their curiosity. All instructional techniques must project a positive affirmation of belief in the abilities of our students. We must show that we care and respect our students and that we want them to succeed. Instructional technique is not separate from content. For example, in teaching proofs one should consider the way in which mathematicians work; perhaps sometimes, instead of presenting proofs, one could have students do proofs themselves. It is important to encourage students to take an active role in their own learning (e.g., through reading assignments, discussion of why material is important, standards for homework). Another important habit to cultivate in students is generalization. One should expect students to solve unfamiliar problems; this may require giving them practice and support.

Students can learn from each other; having them work together can also develop a spirit of camaraderie that helps a course succeed. It can also expose them to multiple methods for solving a problem, thus providing a more robust sense of possible approaches. Besides using the technique of in-class group work, one can get students to work together by explicitly organizing them into homework groups, distributing contact information, requiring them to send questions to a list-serve, or making group work a graded component of the course.

Choice of the means of assessment for grading purposes has an effect on learning. Although timed tests have a place in measuring skill development, many aspects of mathematical learning cannot be effectively measured by timed tests. Individual and group projects, take-home tests, on-line quizzes and assignments, in-class and other oral assessments can be effective means to obtain a rounded view of student achievement and to promote learning of mathematics beyond skill acquisition.

In addition to assessments for grading, it is important to have ongoing mechanisms for gathering information about the state of students' knowledge and understanding in order to teach effectively. There are a multitude of instructional techniques that achieve this, such as calling on students to supply steps in a proof or calculation, quick quizzes on each class's material at the end of each class, requiring students to explain their reasoning in writing and asking students to supply questions by email before each class. Individual instructors will have their own preferences and will be more comfortable and more effective with some methods as opposed to others.

Education research, both that specific to mathematics and outside, provides some basic guide to instructional technique (for example, the powerful effect of teacher expectations). Chapter 9 of the NRC report "Adding It Up" provides a guide. (This report is available from the National Academy Press. See `www.nap.edu`. In the draft available on 15 February, the relevant section begins on page 9–18.)

Instructional choices should be made on the basis of effectiveness, not on complacency and comfort. Instructors should be encouraged to examine their preferences and experiment with new techniques. At the same time, instructors should not be forced to use instructional methods that they believe are unsuited to their skills and preferences. Institutions should provide information on different techniques for mathematics instruction, as well as support for experimentation.

Recruitment

If you want to recruit, teach good courses. We lose a lot of students who come in thinking that they want to be math majors by failing to make real to them the intellectual vitality of mathematics. It is important to assign excellent teachers to the pipeline courses.

Mathematics departments must actively encourage promising students to consider further mathematics courses and ultimately a mathematics major. Instructional techniques that create barriers between students and instructors will hurt recruitment efforts. Instructional techniques must be effective with a broad range of students, not simply with students planning graduate study in mathematics. The profession will pay a dear price if it only encourages the elite to pursue majors in mathematics.

Here are some suggestions for effective student recruitment:

- Put topics and activities into service courses that will attract students to mathematics.
- Present recent developments in mathematics. Students will be excited by material that is interesting, engaging, and intellectually challenging but accessible.
- Use simulations, visualizations and approximations in the presentation of mathematics. Mathematical modeling, which can dramatically illustrate the applicability of mathematics, can draw students into the major.
- Enrich service courses with applications that will both serve students in other departments well and show students the value of mathematics. This possibility is enhanced in calculus courses directed at specific scientific disciplines.
- Promote the idea of mathematics as a profession, with evidence that a major in mathematics is a marketable degree, assistance in obtaining internships, and information about career opportunities in mathematics.
- Seek out students in calculus courses, invite them to be mathematics majors and help them with any necessary paperwork (e.g., by filling out the declaration of major form for them).
- Integrate students into the social as well as the academic life of the department (e.g., by having a departmental lounge accessible to students).
- Employ students in various ways. Jobs, from instructional and tutorial functions to simple clerical duties, can be a powerful draw into the mathematics major. It can greatly add to the students' feelings of involvement in a mathematics community.
- Promote the formation of math clubs, MAA student chapters, Pi Mu Epsilon chapters, Math Circles or other organized mathematics activities.
- Encourage double majors.

WORKSHOP PARTICIPANTS

Joachim Apel, University of Leipzig
Deborah Lowenberg Ball, University of Michigan
William Barker, Bowdoin College
Hyman Bass, Columbia University
Mira Bernstein, Harvard University
Michael Boardman, Pacific University
David Bressoud, Macalester College
Dick Canary, University of Michigan
David Carlton, Stanford University
Jordan Ellenberg, Princeton University
Wade Ellis, West Valley College
Bonnie Gold, Monmouth University
Fred Goodman, University of Iowa
Fernando Gouvea, Colby College
Terry Herdman, Virginia Polytechnic Institute and State University
Roger Howe, Yale University
Thomas Judson, University of Portland
Herbert Kasube, Bradley University
Steven Krantz, Washington University
Jim Lewis, University of Nebraska
Mark MacLean, University of Wisconsin
Gretchen Matthews, Louisiana State University
William McCallum, University of Arizona
John McCleary, Vassar College
Robert Megginson, University of Michigan
Kenneth Millett, University of California, Santa Barbara
Kimberly Pearson, Valparaiso University
Ann Podleski, Harris-Stowe State College
Tom Roby, California State University, Hayward
Stephanie Singer, Haverford College
David Smith, Duke University
Ronald Solomon, Ohio State University
William Velez, University of Arizona
Hung-Hsi Wu, University of California, Berkeley

Physics

CRAFTY Curriculum Foundations Project
Bowdoin College, October 28–31, 1999

Karen Cummings and Guy Emery, Report Editors
Guy Emery and William Barker, Workshop Organizers

Summary

For success in introductory physics, the workshop participants believe that it is most important for students to be able to think operationally within the context of a few fundamental mathematical concepts. While formal theoretical knowledge of the concepts and a tool bag of techniques and computational skills are desirable, the most important factor is that students gain enough active understanding that they are able to think through and solve a wide variety of problems involving the fundamental concepts in a wide variety of contexts. In particular, we present several statements about the general nature of mathematics instruction relevant to the needs of physics students:

1. Conceptual understanding of basic mathematical principles is very important for success in introductory physics. It is more important than esoteric computational skill. However, *basic* computational skill is crucial.
2. Development of problem solving skills is a critical aspect of a mathematics education.
3. Mathematics instruction is worthwhile not only in developing problem-solving skills but also in exposing students to "how a mathematician sees the world." On this point Anthony French of the Massachusetts Institute of Technology says,

   This is a point of view, a rigor, that we value but wouldn't attempt and can't afford in the introductory physics classroom.
4. Introductory mathematics courses should be taught by the best, most skilled teachers; in most cases, that means professors or advanced graduate students.
5. Courses should cover fewer topics and place increased emphasis on increasing the confidence and competence that students have with the most fundamental topics.
6. There are several mathematical topics that are very important for students who will take introductory physics. We list these topics in a table in the next section of this report.
7. Technology should not have a major effect on what mathematics is learned in the first two years. Computers are most helpful for visualization, and for handling problems that are otherwise impracticable. Most instructors of introductory physics are not using symbolic manipulation packages. Spending time in mathematics courses teaching students to use such programs does not directly help in introductory physics courses. However, knowledge of such software is helpful once students enter upper-level physics courses.
8. It is very important to have dialog between the disciplines about what is needed in introductory courses.
9. The impact of mathematics teaching reform on the performance of students in physics courses has not yet made itself felt. However, there is great potential synergism between mathematics education reform

and physics education reform. Ernst Breitenberger made a strong encapsulating statement, which with minor editing may be given as:

The learning of physics depends less directly than one might think on previous learning in mathematics. We just want students who can think. The ability to actively think is the most important thing students need to get from mathematics education.

Narrative

Understanding and Content

What conceptual mathematical principles must students master in the first two years? What broad mathematical topics must students master in the first two years? What priorities exist between these topics?

There were several conceptual mathematical principles that the panel suggested should be mastered in the first two years of mathematics instruction. For example, students should understand that mathematics is the language of science and engineering, and that they must be able to use the language to communicate their knowledge of these other disciplines. They need to be able to translate a problem written in words into the corresponding mathematical statement. Then they must have the computational skill to solve the problem. They must develop a belief in mathematical rules and a comfort with symbols, diagrams, and graphs. They must learn to distinguish between a mathematical concept and techniques for calculation of solutions. (For example, students should know that being able to integrate is quite different from understanding what integration is). They must go beyond "learning rules" to develop understanding. They must acquire "operational" knowledge.

In terms of topical coverage, the panel felt that a solid, deep, functional understanding of the basics was so important that, to make it possible to achieve it, they were ready to reduce exposure to the complexities and intricacies of more difficult ideas and techniques. For example, effective introductory physics instruction requires that students have complete confidence in their ability to understand and calculate simple derivatives and integrals. The more esoteric and complicated topics are not of use in the introductory courses. Furthermore, they are forgotten (and must be re-learned) by the time students reach upper-level courses.

To be successful in undergraduate physics courses, students must understand the *concepts* of function, derivative (as a rate of change and a slope), and integral (as a sum, an area, and an antiderivative). They must also have well-placed confidence in their trigonometric and algebraic skills.

There was significant agreement regarding the topics needed by students taking introductory physics courses. There was also significant agreement about the priorities among these topics. This is summarized in the following table.

Priority	Topics
Highest	Behavior of Simple Functions Derivatives of Simple Functions Integrals of Simple Functions
Higher	Limiting Cases Differential Equations Maximum/Minimum Problems Line and Surface Integrals Vector Operations
High	Series and Sequences Polar and Other Coordinate Systems

By "simple functions" we mean those on a par with polynomials, sine, cosine, tangent, natural logarithm, and exponential functions. "Limiting cases" refer to questions such as "What is the behavior of *e-kt* as *t* gets very large or very small and *k* is either positive or negative?"

What mathematical problem solving skills must students master in the first two years?

The panel judged the development of problem-solving skills to be very important. This was initially expressed as the importance of "teaching students to think." Several panel members thought this was *the* most important outcome of a good mathematics education. The panel was unanimous in suggesting that practice in solving problems should be extensively couched in real-world contexts that are meaningful to the students. Ernst Breitenberger said:

> *I investigated a situation in which physics graduate students were found to be more successful as mathematics teachers than a cohort of mathematics graduate students. The indication was that this was because the physics graduate students used real life examples to teach the mathematics rather than teaching the mathematics without context.*

Several related skills were identified as being important in the development of problem solving abilities. Namely, students should be able to focus a situation into a problem, translate the problem into a mathematical representation, plan a solution, and then execute the plan. Finally, students should be trained to check a solution for reasonableness. Peter Heller, of Brandeis University, adds:

> *Students should get exposure to the general (marvelous) idea that approximations lead to exactitudes.*

What is the desired balance between theoretical understanding and computational skill? How is this balance achieved?

Students need conceptual understanding first, and some comfort in using basic skills; then a deeper approach and more sophisticated skills become meaningful. Computational skill without theoretical understanding is shallow. However, computational skill in several different contexts is usually necessary in development of a more complete understanding of a concept. In discussing theoretical understanding, the panel argues that students should understand the ideas rather than formal proofs. Students should be able to calculate simple cases with great facility. More complicated cases can be left to technology. To achieve the required level of computational skill, most students need significant practice, so some drill-type problems may be required.

Technology

How does technology affect what mathematics should be learned in the first two years?

The ubiquity of information technology has increased the importance of, and need for, training in computer science. The resulting rise in the number of students taking computer science courses indicates a need for a discrete mathematics curriculum that parallels the existing calculus-based curriculum.

Except for that aspect, technology should not fundamentally affect what students learn in their first two years of mathematics instruction. The increased importance of technology to society may have an effect on the topics instructors choose to cover in introductory physics courses. Regardless, the requirements for mathematical preparation of students will not shift significantly, even if additional contemporary technology-related topics are incorporated into introductory physics courses.

Furthermore, the increased prevalence of instructional technologies, e.g., graphing calculators and computers in the classroom, should not fundamentally affect what mathematics students learn in the first two years. Computers offer extraordinary possibilities for enhanced mathematical learning through visualization. Additionally, they relieve the burden of mundane but intricate calculations, and thereby allow instructors to address real world problems that previously were "too messy" to be discussed. However, technology should not drive the curriculum in any way.

The use of computers and symbolic manipulation software in introductory mathematics courses does make possible a desirable shift in emphasis among secondary skills and topics. For example, the memorization of esoteric integration techniques could be displaced by development of skill in prediction of approximate answers and evaluation of whether an answer produced by the computer or calculator is reasonable. Such a shift in emphasis would reinforce changes widely supported by physics education researchers.

Larry Kirkpatrick of Montana State University, former President of the American Association of Physics Teachers, effectively sums up the dominant opinion. He writes:

> *...[T]echnology (in the mathematics curriculum) allows students to spend their time understanding the basic, simple cases, both conceptually and algorithmically. Once students have mastered the simple cases, they can use technology to obtain results for the more complicated cases, much as practitioners or researchers do. When used in this way, technology also better prepares them for entering the work force.*

What mathematical technology skills should students master in the first two years?

There is general agreement among the panelists that some experience with a symbolic manipulation program like *Mathematica* or *Maple* is desirable. However, the point was clearly made by several individuals that these skills are typically not exploited in introductory physics courses. Use of *Maple* or *Mathematica* in an introductory physics course diverts too much of the students' time and attention away from learning the fundamental physics. Hence, development of that kind of skill is only of significant value once students enter junior and senior level physics courses. At that point, such skill is easily exploited to facilitate further learning.

Specifically, students should become skilled in basic computational techniques. For example, they should become very comfortable with the use of symbols and naming of quantities and variables. They get important experience with this when they work on realistic (real world) problems. They should be fluent in the use and meaning of the derivative, and should be able to formulate and solve simple differential equations. They should also be able to use a calculator or computer to solve simultaneous algebraic equations. Peter Heller gave a compelling alternative opinion:

> *Spread sheets are by far the best medium (for teaching with technology) since data, text and graphics are all visible at once, and since the techniques are easily learned and useful for numerical approximations.*

Instructional Techniques

What instructional methods best develop the mathematical comprehension needed for your discipline?

While the panel did not have enough specific knowledge of the methods typically used in mathematics instruction to answer this question very fully, there were a number of general responses. The methods chosen for use should be based on research about the ways students learn and the ways they don't. In general, any method that teaches conceptual understanding as well as useful algorithms was acceptable. There was no objection to drill practice as a way to develop competency in using simple algorithms.

Peter Heller points out that his research, which is based on interviews with hundreds of students, indicates that the traditionally used methods of mathematical instruction fail to reach the vast majority of students. In a paper to be published he writes:

> *When it comes to conceptual understanding of the basics, the traditional methods of calculus instruction did not seem to have reached most of the students I interviewed. Let me give an example. All the students could say that the derivative of the function $y = f(x) = x^2$ is $f'(x) = 2x$. But when asked: "Can you use that to estimate the square of 10.1 knowing that 102 = 100",*

the vast majority didn't know where to begin, said that y was 2Δx, or that Δy was 2Δx (thereby relating a finite quantity with a small-increment). When I did this sort of thing five years ago for the natural logarithm function I even saw Δx going into the denominator! All in all, they didn't seem to appreciate that calculus is fundamentally about numbers.

What guidance does educational research provide concerning mathematical training in your discipline?

A great deal. However, education research results must always be adapted to the local constraints. This is typically a subtle process. Engineering is a good model for how to apply a general principle to a local situation even with imperfect knowledge.

— Kenneth Heller, Professor of Physics, University of Minnesota and former Chair of the American Physical Society's Forum on Education.

Education research indicates basic methods that generally lead to increased pedagogical efficacy regardless of discipline. Such research indicates, for example, that students must have hands-on experience with the material, that examples should be "real world," that there should be a focus on understanding the fundamental concepts in several different representations (natural language, graphical, symbolic, . . .), and that fewer topics should be covered and those topics included should be covered in greater depth.

Instructional Interconnections

What impact does mathematics education reform have on instruction in your discipline?

To date, the impact of mathematics education reform on physics instruction has been small. However, the panel feels that the potential impact is great. We hope mathematics education reform will reinforce similar efforts in physics education reform, since these movements stem from a common philosophical viewpoint and hence advocate similar approaches to instruction.

If the result of mathematics education reform is that students have a deeper understanding of fundamental mathematical concepts and a heightened ability to do basic mathematical manipulations, the impact on physics instruction will be large and positive. Improvements in conceptual understanding of mathematical ideas will greatly facilitate conceptual learning in physics. Improved learning (understanding) of simple algorithms by spending more time on them and less time on more esoteric algorithms would be very useful. However, it is also clear that unless the reform is done carefully, the removal of important topics could have a significant detrimental effect on physics instruction.

Perhaps Stanley Haan, Professor and Chair, Department of Physics and Astronomy at Calvin College, made the most thoughtful comment related to the potential impact of mathematics education reform. He said:

(Mathematics education reform) will have an enormous impact on physics if it can help to change students' conception of what 'good teaching' ought to be like.

How should education reform in your discipline affect mathematics instruction?

On this topic, Barry Holstein, Professor of Physics, University of Massachusetts, wrote:

The essence of education reform in physics is in creating methods of instruction which are more effective in conveying knowledge that students retain. Such methods are universal and certainly have relevance to instruction in mathematics (and other subjects).

The panel agreed. The basic ideas about "best practice" in instruction are not likely to be highly discipline specific. For example, Peter Heller suggests that statements made within the physics education community (like "start from where the student is, not where the teacher thinks the student is," and "really try to understand what is in the mind of the student") are directly applicable to mathematics education as well. In addition, Kenneth Heller argues that physics education reformers' suggestions that students should get extensive practice with writing reasons for their answers, with communicating their thoughts on procedures, with solving real problems (where the path to the answer is not known by the student at the beginning), with applying their knowledge in a context meaningful to them, with making connection to other domains of their knowledge, and with working effectively in cooperating groups, are directly applicable to mathematics instruction.

Consistency of approach in mathematics and physics, and mutual reinforcement, may be quite important in helping each student achieve his or her own potential. Furthermore, physics and mathematics education reforms should be coordinated so they can be supported by both disciplines, and so that both disciplines can benefit from research findings.

WORKSHOP PARTICIPANTS

(See the Appendix for detailed biographies.)

Ernst Breitenberger, Professor of Physics, Emeritus, Ohio University
Karen Cummings, Assistant Professor of Physics, Rensselaer Polytechnic Institute
Guy Emery, Professor of Physics, Emeritus, Bowdoin College
Anthony French, Professor of Physics, Emeritus, Massachusetts Institute of Technology
Stanley Haan, Professor and Chair of Physics and Astronomy, Calvin College
Randal Harrington, Physics Teacher, San Jose High School
Kenneth Heller. Professor of Physics, University of Minnesota
Peter Heller, Professor of Physics, Brandeis University
Barry Holstein, Professor of Physics, University of Massachusetts, Amherst
Larry Kirkpatrick, Professor of Physics, Montana State University
Malgorzata Zielinska-Pfabe, Professor of Physics, Smith College

Mathematics Participants

Thomas Banchoff, Professor of Mathematics, Brown University
William Barker, Professor of Mathematics, Bowdoin College
Thomas Berger, Professor of Mathematics, Colby College
Susan Ganter, Associate Professor of Mathematical Sciences, Clemson University
Deborah Hughes Hallett, Professor of Mathematics, University of Arizona
Harvey Keynes, Professor of Mathematics, University of Minnesota
William McCallum, Professor of Mathematics, University of Arizona
Donald Small, Professor of Mathematics, U.S. Military Academy, West Point

ACKNOWLEDGEMENTS

The members of the panel would like to thank their colleagues in the Mathematical Association of America for providing this opportunity to consider and make recommendations on a topic of importance; William Barker, the chair of CRAFTY; Allen Tucker and Katharine Billings for helping with the planning and operation; those mathematicians who worked directly with us during the workshop: Tom Banchoff, William Barker, Tom Berger, Susan Ganter, Deborah Hughes Hallett, Harvey Keynes, William McCallum, and Don Small; the computer scientists who took part in the concurrent CS program; and Bowdoin College for sponsoring the Workshop on the occasion of the Rededication of the renovated Searles Science Building.

APPENDIX: Biographies of the Workshop Participants

Ernst Breitenberger earned a Dr. phil. in real functions under Johann Radon in Vienna, 1950. Thereafter, he earned his keep on four continents mostly in physics and its penumbra. He has worked in the areas of high-voltage engineering; low-energy nuclear experiments (Ph.D. Cambridge, 1956); theoretical physics; statistics and stochastic processes; operations research; science history and biography. He was Professor of Physics at Ohio University in Athens, Ohio, from 1963; becoming emeritus in 1994. His concern about the teaching of mathematics to non-mathematicians led to a semi-quantitative study and an article entitled: "The Mathematical Knowledge of Physics Graduates," American Journal of Physics 60, 318–323 (April 1992).

Karen Cummings received her Ph.D. from the University at Albany, State University of New York, in 1996. Her dissertation research involved the use of nuclear reaction analysis and scattering techniques for materials analysis. Dr. Cummings joined the faculty at Rensselaer Polytechnic Institute in 1997 following a one year sabbatical replacement position at Skidmore College. She is currently the Edward Hamilton Clinical Assistant Professor of Physics at Rensselaer. Professor Cummings was awarded the Hamilton title in recognition of her innovative work in undergraduate physics education. In addition to her teaching, she does federally funded research on the learning and teaching of introductory university physics. Her current funded projects include the development of a tool for assessment of student problem-solving ability, curriculum development for introductory courses and using assessment as a guide to increased learning outcomes in first year physics courses. Professor Cummings is the course supervisor for the Studio Physics I course at RPI, which involves eight faculty members, ten teaching assistants and 600 students per semester in an active engagement style course. She is currently working on a revision of the widely used physics textbook, *Fundamentals of Physics.* The revised text will incorporate modern pedagogical approaches and research in physics education. It will be titled *Fundamentals of Physics* by Halliday, Resnick, Walker and Cummings and is scheduled for release in the Fall of 2001.

Guy Emery studied at Bowdoin (A.B. 1953) and Harvard (Ph.D. 1959). He was at Brookhaven National Laboratory (1959–1966), then moved to Indiana University where he became Professor of Physics. He worked at the Indiana University Cyclotron Facility, where he was Liaison Officer for the Users Group (1970–1979) and Associate Director for Research (1972–1979). Professor Emery returned to Bowdoin in 1988 as Department Chair and became emeritus in 1998. He has also been a visitor at the State University of New York at Stony Brook, Groningen, and Osaka. His research papers are in the areas of nuclear spectroscopy and nuclear structure, chemical and solid-state effects on nuclear processes, nuclear reactions with large momentum transfer, and production of pions near threshold. In recent years he has been actively pursuing the history of twentieth-century physics.

Anthony French is Professor of Physics, Emeritus, at Massachusetts Institute of Technology. He received his B.A. (1942) and Ph.D. (1948) at Cambridge University, and was demonstrator and Lecturer (1948–1955) at the Cavendish Laboratory where he did research in nuclear physics. He was Professor of Physics at the University of South Carolina (1955–1962, Dept. Head 1956–1962) and Visiting Professor at MIT (1962–1964) while working on a physics curriculum development project for undergraduate students. He was Full Professor at MIT from 1964–1991. Dr. French has authored textbooks, student experiments, demonstrations, and films. His textbooks include: *Special Relativity, Newtonian Mechanics, Vibrations & Waves, Introduction to Quantum Physics* (with E. F. Taylor). He is editor of *Einstein: A Centenary Volume* (1979) and co-editor of *Niels Bohr: A Centenary Volume* (1985). Professor French was awarded the Institute of Physics (UK) Bragg Medal in 1988, the American Association of Physics Teachers' Oersted Medal in 1989 and the Melba Phillips Award 1993. He was President of the AAPT from 1985 to 1986. His chief current professional interests are the subject matter of undergraduate physics and the history of physics.

Stanley Haan is Professor and Chair in the Department of Physics and Astronomy at Calvin College in Grand Rapids, Michigan. He received his Ph.D. in 1983 from Colorado. His professional interests are in theoretical atomic physics, especially as relating to photorecombination and strong-field photoionization, and in science education, especially as relating to elementary school science. Professor Haan has thirty-two physics research publications, including eight with 13 different undergraduate research assistants as co-authors. He has been principal investigator on several National Science Foundation grants for physics research and is co-principal investigator on a NSF grant to develop a course in scientific analysis for elementary-education students. He is actively involved in course development for elementary education students and has three science-education publications. Professor Haan has led numerous workshops for elementary school teachers.

Randal Harrington is the founder and former Director of the Laboratory for Research in Physics Education at the University of Maine, is currently teaching High School Physics in San Jose, California. He received his Ph.D. in Physics from the University of Washington in 1995 and has over 20 years of teaching experience. Dr. Harrington is active in the American Association of Physics Teachers where he has served as the Chair of the Committee on Research in Physics Education. In the past, he has received funding from the National Science Foundation, served on national physics test construction committees, and has organized and run numerous workshops and seminars on teaching physics and integrating technology into the science classroom. He has also served as a content consultant for both Microsoft and the education division of WGBH, and has worked as either a consultant or an external evaluator on numerous physics reform projects. In 1997, Dr. Harrington was awarded a Higher Education SEED's Fellowship from the Maine Mathematics and Science Alliance and the Maine Department of Education for his work with preservice teachers and for reforming the introductory physics course at the University of Maine. Dr. Harrington has co-authored several curricula including units on electric charge, magnetism, radiation and radioactivity.

Kenneth Heller received his B.A. from the University of California (1965) and his Ph.D. in physics from the University of Washington in Seattle in 1973. He became research associate and instructor at the University of Michigan in 1972 and joined the faculty of the University of Minnesota in 1978, becoming Professor of Physics in 1986. Professor Heller was visiting professor at the University of Utah in 1985 and a member of the board of trustees of the Universities Research Association (1985–1988). He has been principal investigator in high energy physics at Minnesota since 1989. His experimental particle physics research has included studies of quark dynamics from strong interactions of hadrons, quark confinement from magnetic moments of baryons and their weak decay properties, and muons from high energy interactions. Professor Heller was Chair of the American Physical Society's Forum on Education at the time of the Worskhop.

Peter Heller was educated at MIT and received his Ph.D. from Harvard University in 1963. Following postdoctoral research at MIT, he joined the faculty at Brandeis (Asst. Prof. 1966, Prof. 1974). His experimental researches on phase transitions and critical phenomena helped launch the modern era in that subject. He also has a long-standing interest in mathematics and physics education going back to 1958, when he worked in an NSF-sponsored program for the training of high school teachers of science and mathematics. Since 1982 he has devoted all his time to developing novel physics and mathematics teaching experiments and approaches at levels ranging from high school to graduate school. This has led to several well-known papers and participation with several groups including the editorial board of the American Journal of Physics. At Brandeis, Professor Heller has been a recipient of the university-wide Louis Dembitz Brandeis Prize for Excellence in Teaching. Within the past year he has been cited by graduating seniors as one of the "most influential teachers" with respect to contributions to student intellectual development. Analysis of student comments shows this to be largely due to his efforts in the teaching of mathematics within physics courses. This aspect has involved more than a thousand hours of interviews and Socratic teaching sessions with individual students.

Barry Holstein was educated at Carnegie Tech (B.S. 1965, M.S. 1967) and Carnegie-Mellon University (Ph.D. in physics, 1969). He became instructor in physics at Princeton in 1969, joined the faculty at the University of Massachusetts in 1971, and became full professor in 1979. He has been visitor at Princeton (fellow 1975–76, Prof. 1985) and program officer in theoretical physics at the National Science Foundation (1977–1979). He is a theoretical physicist who has specialized in weak interactions, including nonleptonic weak processes, weak decays of nuclei, and chiral symmetry. He has published several pedagogical articles in the American Journal of Physics, most recently "The Neutrino," with Wick Haxton, 68, 14–31 (January 2000).

Larry Kirkpatrick is Professor of Physics at Montana State University. He received his Ph.D. in experimental high energy physics from MIT in 1968. After five years at the University of Washington, he moved to Montana State University as a teaching specialist in physics with the responsibility for training teachers and improving the teaching of physics at all levels. In addition to his extensive activities in state and national physics and science teaching organizations, Professor Kirkpatrick served for eight years as a coach of the US Physics Team. He also co-writes the physics contest column and serves as the field editor for physics for *Quantum Magazine*, a joint US-USSR publication for bright students in mathematics and physics. Professor Kirkpatrick has been recognized for his physics teaching (Phi Kappa Phi Outstanding Teaching Award and the Burlington Northern Foundation Faculty Achievement Award for Outstanding Teaching) and his service to the physics teaching profession (Distinguished Service Citation from the American Association of Physics Teachers and a Merit Citation from the Montana Science Teachers Association). He was President of the American Association of Physics Teachers at the time of the Workshop, and is a fellow of the American Physical Society and a member of the Governing Board of the American Institute of Physics. In addition to his 14 research publications in high energy physics, he has published a textbook (*Physics: A World View*) now in its third edition, seven articles on physics education, and numerous reviews and news articles. In his service to the physics teaching profession and the general public, Professor Kirkpatrick has given more than 300 workshops and presentations.

Malgorzata Zielinska-Pfabe received her M.Sc. degree in theoretical physics from the University of Warsaw, Poland. She received her Ph.D. in nuclear physics (1969) from the Institute of Nuclear Research in Warsaw, where she worked as a research fellow until 1978. She was a visiting researcher at Rensselaer Polytechnic Institute from 1978 to 1982. In 1982 she joined the faculty at Smith College, and became a full professor in 1984. Professor Zielinska-Pfabe was awarded the first Smith College Senior Faculty Teaching Award in 1985 and is a fellow of the American Physical Society. In 1992 she was appointed to the Sophia Smith Chair in Physics. Professor Zielinska-Pfabe's field of research is theoretical heavy ion physics. She has published over 50 papers in refereed professional journal in the US, Europe, and Australia, and authored over 50 professional abstracts and conference proceedings.

Statistics

CRAFTY Curriculum Foundations Project
Grinnell College, October 28–31, 1999 and October 12–15, 2000

Tom Moore, Roxy Peck and Allan Rossman, Report Editors and Workshop Organizers

Summary

This report addresses the role of undergraduate mathematics in preparing students to study statistics and the role of statistics in the undergraduate mathematics curriculum. Statistics is a partner discipline as well as a client discipline of mathematics. By this we mean that statistics is a part of the mathematical sciences and should be represented within the curriculum as addressed by the MAA; at most undergraduate institutions, there is no separate statistics department and so responsibility for statistics offerings typically falls to the mathematics department.

The two highest priority needs of statistics from the mathematics curriculum are to:

1. Develop skills and habits of mind for problem solving and for generalization. Such development toward independent learning is deemed more important than coverage of any specific content area.
2. Focus on conceptual understanding of key ideas of calculus and linear algebra, including function, derivative, integral, approximation, and transformation.

The following recommendations are necessary to achieving the two recommendations above. They are listed in decreasing order of importance as determined by the focus group, but all are considered key by both focus group and workshop participants.

1. Emphasize multiple representations of mathematical objects and multiple approaches to problem solving, including graphical, numerical, analytical, and verbal.
2. Instruction should be learner-centered and address students' different learning styles by employing multiple pedagogies.
3. Insist that students communicate in writing and learn to read algebra for meaning.
4. Use real, engaging applications through which students can learn to draw connections between the language of mathematics and the context of the application.
5. Instill appreciation of the power of technology and develop skills necessary to use appropriate technology to solve problems, to develop understanding, and to explore concepts.
6. Align assessment strategies with instructional goals.

In looking at this set of recommendations, the workshop participants observed a strong consistency between them and what we consider to be principles of calculus reform. It was not our intention to take sides in what has, at times, been a source of contention within the mathematics community. Rather, this consistency was a natural by-product of our deliberations.

A second set of recommendations was developed to address statistics' role as part of the mathematical sciences:

1. We endorse the 1991 CUPM recommendation that every mathematical sciences major should study data analysis and statistics.
2. We should advocate relaxing the assumption that the first course in statistics for majors must have the calculus pre-requisite stated by the 1991 CUPM recommendation.
3. We can create a wider acceptance of this recommendation by providing compelling arguments that this need is even greater now than a decade ago and by offering examples of a diverse set of successful courses that address this goal. (We include a diverse set of short course descriptions in the appendices.)
4. We endorse the 1992 recommendations of the ASA/MAA Committee on Undergraduate Statistics for teaching introductory courses in statistics— to emphasize statistical thinking through active learning, with more data and concepts, less theory and fewer recipes.
5. We encourage those responsible for the mathematical needs of students majoring in client disciplines to recognize in their curricular offerings and educational requirements that many of these students would be well served by a statistics course that teaches them how to deal with data-oriented problems in their discipline.
6. We encourage those responsible for the general education requirements in quantitative reasoning to recognize in their curricular offerings and educational requirements that many students would be well served by a modern statistics course that meets the 1992 recommendations of the ASA/MAA Committee on Undergraduate Statistics.

Narrative

Understanding and Content

We felt the tension between content and process, between covering lots of topics and making students think and learn how to solve problems. A list of principles emerged from an early discussion and kept recurring in our conversations. (For the purpose of this section, we use the term 'mathematics' to exclude statistics and consider the issue of what mathematical knowledge and skills we would like statistics students to develop, particularly in the first two years of an undergraduate program.) This list became known as George's list (named for the list's founder, George Cobb).

Desired outcomes of mathematics courses

1. Process of abstraction
2. Ability to distinguish the relevant from the irrelevant
3. Ability to go from special cases to generalizations
4. Ability to test and verify conjectures
5. Ability to use abstract ideas in applied situations
6. Ability to make abstract connections
7. Ability to explain oneself logically

We think of mathematical content as a means to these ends. We believe that the outcomes on this list are not attained through one course or even through several courses, but are achieved incrementally. Attention should be paid throughout the undergraduate curriculum (and before), so that content does not drive out the need to expose students to mathematical experiences that will further these outcomes. These outcomes enable more successful the learning and application of statistics, and we can live with variations in content choices.

Still, there are some fundamental ideas from the traditional first two years that would hinder our teaching were they to be absent from our students' backgrounds. This minimal content set includes key ideas of calculus including the concept of function, derivative as rate of change, integral as accumulation, and approximation. Also included are key ideas from linear algebra including linear transformation, projections, and visualizing Euclidean n-space. Basic computational skill and student facility with the interplay among graphical, numerical, algebraic, and verbal representations is also essential.

Technology

With regard to the question of how technology affects what mathematics should be learned in the first two years, the group listed three primary areas:

1. Technology enables more emphasis to be placed on developing students' skills of problem solving. This includes encouraging students to establish the mindset that multiple problem-solving strategies (graphical, numerical, algebraic) are possible. Technology also permits students to analyze "real" applications as opposed to contrived ones.
2. Technology can facilitate students' development of conceptual understanding.Visualization is one avenue through which this can occur, for example with concepts such as approximation by Taylor series. The use of dynamic graphics can help students to understand concepts of calculus. The ability of technology to handle some symbolic manipulations can sometimes allow students to focus their attention on understanding concepts.
3. Technology can promote students' exploration of and experimentation with mathematical ideas. For example, students can be encouraged to ask "what if?" questions, to posit conjectures, to answer them, and to use technology to investigate, revise, and refine their predictions. Specific examples include studying the effects of manipulating parameters on classes of functions and fitting functional models to data.

In terms of specific content related to the technology to be taught and learned, the group cited examples which include numerical issues associated with approximations and round-off errors and also applications of linear algebra such as least squares estimation.

Concerning the question of what mathematical technological skills students should master in the first two years, the group listed both general and specific skills. General skills include knowledge of the usefulness of technology, willingness to use technology (without being specifically told to do so), and understanding the importance of choosing appropriate technology tools, including an appreciation of the strengths and weaknesses of different tools. Specific skills include the ability to use technology to:

1. Graph functions
2. Numerically evaluate functions
3. Solve equations, calculate derivatives and integrals (including multivariable functions)
4. Implement algorithms
5. Perform symbolic manipulations
6. Perform matrix operations

Instructional Interconnections and Techniques

For all students:

Addressing questions of interconnections of instructional ideas between statistics and mathematics communities, the group commented that there are many common ideas between statistics education reform and calculus reform efforts. These include emphases on conceptual understanding, active learning, real applications, and the use of technology. One difference, though, is that there appears to be a much broader consensus among statisticians on these principles of statistics education reform than there is among mathematicians with regard to calculus reform. Most of the work in statistics education reform has been directed at the introductory service course and not at courses that have a mathematics prerequisite.

The group also noted that these efforts can support each other. For example, students who learn about fitting models to data in a calculus class can build on that knowledge in a statistics course. Another example is that mathematical statistics courses can build on students' abilities to use technology to perform calculations involving multivariable calculus in the study of multivariate probability distributions.

Addressing the need for collaboration between the mathematics and statistics communities, it was noted that several mechanisms for such collaboration already exist. The MAA and ASA have had an active joint committee devoted to issues of undergraduate statistics for many years, and the MAA has recently established a Special Interest Group devoted to statistics education. These groups need to continue to ensure that communication flows freely between the two groups. The importance of teacher preparation was mentioned as an area in which this communication and collaboration are especially important.

Some other issues discussed with regard to these questions of instructional techniques and connections were that:

1. Multiple techniques of teaching and learning must be employed because of multiple goals and multiple audiences.
2. Students must learn to learn from different instructional techniques.
3. Students must learn to see connections for themselves, for instance between the language of mathematics and the context of an application.
4. Writing should be emphasized as a learning tool for making these connections as well as for the sake of clear communication.
5. Algebra has a role in that students should learn to read algebra for meaning (for example, in the equation for calculating a correlation coefficient based on z-scores of the two variables).
6. Students should learn about the importance of units of measurement. For example, students should understand that the slope and intercept of a least squares line for predicting birth weight from weeks of gestation have different units.

The following passage appears on page 34 of *Everybody Counts* [National Academy Press, 1989]:

> *Mathematical Science is a term that refers to disciplines that are inherently mathematical (for example, statistics, logic, actuarial science), not to the many natural sciences (for example, physics) that employ mathematics extensively. For economy of language, the word 'mathematics' is often used these days as a synonym for 'mathematical sciences', as the term 'science' is often used as a summary term for mathematics, science, engineering, and technology.*

This passage points out the difference between client disciplines (such as physics, biology, and engineering) and what we are calling a partner discipline, namely statistics. Most institutions do not have a separate department of statistics, and we feel that for such institutions the mathematics department should take the responsibility for housing the discipline of statistics. That is, these departments must be mathematical sciences departments. For this reason, the workshop considered the proper place of statistics within a mathematical sciences curriculum and addressed the following two sets of questions:

1. What (if any) statistical concepts and methods are important for all students majoring in a mathematical science, particularly in the first two years? What statistical concepts are essential? What statistical skills are essential? What features should characterize mathematics majors' study of statistics? What role should technology play in this statistics instruction? What impact does reform in pedagogical methods have?
2. What statistical concepts and methods are important for students in quantitative disciplines, particularly in the first two years?

For students majoring in a mathematical science:

The 1991 CUPM Report said:

> *Every mathematical sciences major should include at least one semester of study of probability and statistics at a level which uses a calculus prerequisite. The major focus of this course should be on data and on the skills and mathematical tools motivated by problems of collecting and analyzing data. Any statistics course taught now should use a nationally available software package.*

The workshop considered the question of requirements for the mathematics (mathematical sciences) major. Should every major take a data-centered statistics course? If so, what should it look like?

The 1991 CUPM recommendation on statistics has been widely ignored. One institution that has taken it seriously is Gustavus Adolphus College, which instituted a breadth requirement for mathematics majors that is satisfied by either statistics or computer science. A description of Gustavus Adolphus's course is given in Appendix A. The CRAFTY Statistics Workshop affirms the 1991 CUPM recommendation concerning statistics. The arguments for mathematics majors to gain substantial experience working with issues of data and chance are even more compelling now than a decade ago:

1. Data analysis plays a crucial role in many aspects of academic, professional, and personal life.
2. The job market for mathematics majors is largely in fields (e.g., business) that use data.
3. Future teachers will need knowledge of statistics and data analysis to be current with the new NCTM Standards and with the new and very popular AP statistics course.
4. The study of statistics provides an opportunity for students to gain frequent experience with the interplay between abstraction and context, which we regard as critical for mathematical science students to master.

Requiring a statistics course is good for it can also help the mathematics department in that it broadens the appeal of the department to students and potential majors. The group felt strongly, however, that we should not mandate a calculus prerequisite for the first statistics course. While it is certainly preferable for mathematically inclined students to take a statistics course in which they make use of their mathematical knowledge and ability, not all institutions are able to offer a data-centered statistics course with a calculus prerequisite. The group believes that this requirement may have contributed to the minimal impact of the recommendation and argues that a statistics course lacking a calculus prerequisite can still be an important educational experience for a mathematics major.

We recommend that the course should emphasize the collection and analysis of real data, to show students in this first exposure what the discipline of statistics is primarily about. We recommend that this course adhere to the fundamental principles articulated by Cobb (1992) on behalf of the ASA/MAA joint committee:

1. More data and concepts, fewer derivations and recipes; automate calculations using a modern statistical package
2. Emphasize statistical thinking: the omnipresence of variability and the importance of data production
3. Foster active learning: student projects, group work, activities, writing, oral presentations

A required course for all mathematical science majors should embody these principles, which have the potential of presenting an authentic view of statistics. These broad principles leave room for a variety of first statistics courses for students, and in the appendices we have included descriptions of several examples that are diverse and suggest the possibilities. Courses include:

1. Carolyn Dobler's introductory course at Gustavus Adolphus (Appendix A)
2. A first statistics course in experimental design developed by George Cobb at Mt. Holyoke (Appendix B)
3. A new archaeometrics course, developed by Don Bentley at Pomona (Appendix C)

4. A new data analysis course at the post-calculus level designed by Rossman, Chance, and Ballman (Appendix D)
5. A case studies based mathematical statistics course developed by Deb Nolan and Terry Speed at Berkeley (Appendix E)
6. A time series course developed by Robin Lock of St. Lawrence (Appendix F)
7. A course emphasizing Bayesian statistics developed and taught at Duke University by Dalene Stangl and colleagues (Appendix G)

The course descriptions make it clear that the three principles articulated by the Focus Group are central to those courses and at the heart of each is the focus on data.

Exposing students to such a statistics course as undergraduates allows them to see an area of mathematics that they are probably unfamiliar with (although this may become less true as new NCTM standards are implemented) and this broadens the appeal of the mathematics department to students. Notice that most of the examples described in the appendices are courses that are at the 100- or 200-level, so that early exposure enhances the chances that students will want to take more mathematics or statistics, which should improve departmental enrollments.

For students majoring in quantitative disciplines other than mathematics:

We make a distinction between subjects whose content is often mathematical (engineering, physics, genetics, chemistry, economics, environmental science) and subjects that use data a lot (psychology, sociology, business, biology). Non-majors can also benefit from courses described in the previous sections, especially the more quantitatively ambitious ones. Indeed, most of these courses can be taken before students have declared a choice of major. At the same time, we feel a need for courses that serve the direct needs of client disciplines for students who are less quantitatively ambitious and that serve the general educational needs of the institution. As mentioned earlier in this report, the general education introductory statistics course has been described at length in the literature.

A good survey article on the current state of the general education course comes from a position paper presented in August of 2000 at the ASA's Undergraduate Statistics Education Initiative. This article by Garfield, Hogg, Schau, and Whittinghill (2002) has been published in the Journal of Statistics Education. The paper is "Best Practices in Introductory Statistics", by Joan Garfield, Bob Hogg, Candace Schau, and Dex Whittinghill and is available on line at:

`www.amstat.org/meetings/jsm/2000/usei/usei_1st.PDF`

For students who will take only one college course in quantitative or mathematical reasoning, most will be better served, we think, by a course like the one described in this paper than by a college algebra or precalculus course. Many of these students will find careers that require an appreciation and facility with data and all will be citizens in a world where statistical information is pervasive.

REFERENCES

CUPM (1992), "The Undergraduate Major in the Mathematical Sciences: A Report of The Committee on the Undergraduate Program in Mathematics," *Heeding the Call for Change*, L. A. Steen, editor. The Mathematical Association of America, 1992, 228–247.

Cobb, George (1992), "Teaching Statistics," in *Heeding the Call for Change: Suggestions for Curricular Action,* Lynn A. Steen, ed., The Mathematical Association of America, pages 3–43.

Garfield, Joan, Hogg, Bob, Schau, Candace and Whittinghill, Dex (20020), "First Courses in Statistical Science: The Status of Educational Reform Efforts," Journal of Statistics Education, Volume 10, Number 2, `www.amstat.org/meetings/jsm/2000/usei/usei_1st.PDF`.

MSEB (Mathematical Sciences Education Board) (1989), *Everybody Counts: A Report to the Nation on the Future of Mathematics Education,* National Academy Press, Washington, 1989.

WORKSHOP PARTICIPANTS

Statistics Participants

Mary Batcher, Statistician, Ernst and Young
Beth Chance, Associate Professor of Statistics, California Polytechnic State University
Veronica Czitrom, Statistician, Bell Laboratories
William Duckworth, Assistant Professor of Statistics, Iowa State University
Dave Fluharty, Statistician, ALCOA
Michael Larsen, Senior Lecturer of Statistics, University of Chicago
Carl Lee, Professor of Statistics, Central Michigan University
Russ Lenth, Associate Professor of Statistics, University of Iowa
Deborah Nolan, Professor of Statistics, University of California-Berkeley
Roxy Peck, Professor of Statistics, California Polytechnic State University
Allan Rossman, Associate Professor of Statistics, California Polytechnic State University
Robert Starbuck, Statistician, Amerst Wyeth

Mathematics Participants

Jack Bookman , Associate Professor of the Practice of Mathematics, Duke University
Ann Cannon, Associate Professor of Mathematics, Cornell College
George Cobb, Professor of Mathematics and Statistics, Mt. Holyoke College
Carolyn Dobler, Associate Professor of Mathematics, Gustavus Adolphus College
Marlin Eby, Associate Professor of Mathematics, Messiah College
Deborah Hughes Hallett, Professor of Mathematics, University of Arizona
Steve Horton, Associate Professor of Mathematics, United States Military Academy
Michael Kahn, Associate Professor of Mathematics, St. Olaf College
Tom Moore, Professor of Mathematics, Grinnell College
Mary Parker, Professor of Mathematics, Austin Community College
Harriet Pollatsek, Professor of Mathematics, Mt. Holyoke College
Jon Scott, Professor of Mathematics, Montgomery College University
Kay Somers, Professor of Mathematics, Moravian College
Dalene Stangl, Professor of the Practice of Statistics, Duke University
Ann Watkins, Professor of Mathematics, California State University, Northridge

ACKNOWLEDGEMENTS

We would like to thank Grinnell College for sponsoring the workshop and the American Statistical Association for financial support.

APPENDIX A: Statistics Requirements at Gustavus Adolphus College

(Carolyn Pillers Dobler, Dept. of Mathematics and Computer Science, Gustavus Adolphus College)

In 1993, the Department of Mathematics and Computer Science at Gustavus Adolphus College revised the mathematics major, following the 1991 CUPM report on the undergraduate major. The revised major consists of a basic core (Calculus I, II, III, Linear Algebra, Theory of Calculus), a breadth requirement (Introduction to Computer Science, Introduction to Statistics), a classical core (one course from complex analysis, real analysis, or modern algebra), a depth requirement (an upper level, two-semester sequence), an applied course, and either a capstone course or senior oral exam.

The 1991 CUPM report specifies a course in probability and statistics having a calculus prerequisite, but with the major focus on data, including data collection, data description, and data analysis. In the initial discussions, the department hoped to create a new course, designed specifically for mathematics majors, but staffing restrictions precluded this. Much discussion ensued over which existing course would be appropriate to meet the CUPM guidelines. The upper-level probability and mathematical statistics course was too theoretical, whereas the lower-level introductory statistics course might not be rigorous enough. One option would be to include a data analysis component in the theoretical course, which had been attempted unsuccessfully in the past. Another option would be to make the appropriate calculus connections in the introductory course. Further, the department considered the secondary education mathematics majors. Teacher licensure in the State of Minnesota requires secondary mathematics teacher to have a course in probability or statistics. The department felt that the data-oriented course was much more appropriate for the future teachers, particularly since many may eventually be teaching statistics in a high school. The department decided to include the lower-level introduction to statistics course (with a Calculus I prerequisite) as the required course in the mathematics major.

The course, Introduction to Statistics, is taught each semester to about thirty students, most of whom are mathematics or science majors. The text is Introduction to the Practice of Statistics by David S. Moore and George P. McCabe. Although the text does not specifically use calculus, through supplementation, connections to calculus are made through the topics of the normal distribution, least-squares regression, and probability. The course follows 1991 Statistics Focus Group recommendations to emphasize statistical thinking, to include more data and concepts (less theory, fewer recipes), and to foster active learning. The course is data-driven, emphasizing analysis and interpretation. Students gain experience in technical writing through a semester-long group project and an individual data analysis project. Although there is no formal (weekly) laboratory component, computer lab exercises are interspersed throughout the semester as appropriate.

Mathematics majors typically take the course in their sophomore year, often when they are also taking a more proof-oriented course such as Linear Algebra or Theory of Calculus. Although the student may be taking two mathematics courses in a semester, a balance is achieved between theory and application.

Overall, the department believes the requirement of introductory statistics has been beneficial to most mathematics majors, particularly to those in secondary education. In addition, there has been an increase in the number of students who have selected the upper-level probability and mathematical statistics sequence as their depth requirement after being exposed to statistics in the introductory course. Because most of the students in the upper-level sequence have had the introductory course, the theoretical material can be connected to the applications with which the students are familiar.

APPENDIX B: An Introductory Course that Emphasizes Design of Experiments

(George W. Cobb, Department of Mathematics and Statistics, Mount Holyoke College)

The course described here provides an applied introduction to the design and analysis of experiments for students with no previous background in statistics. At Mount Holyoke, a liberal arts college for 1800 women undergraduates, the course has served as an alternative first course in statistics for our few statistics majors and some of the many students majoring in biology, psychology, and environmental science. Although no statistics or calculus is required or assumed of students who take the course, we do require a semester of (any) 100-level work in mathematics or statistics, and accordingly, the course is listed as 200-level. The next few paragraphs lay out the goals and sketch the approach of the course. Voluminous additional details can be found in a book *An Introduction to the Design and Analysis of Experiments* (Springer, 1997), which evolved, sometimes painfully, over the first 15 years the course was taught at Mount Holyoke.

Goals

Successful students learn to:

- to choose sound and suitable design structures;
- to recognize the structure of any balanced design built from crossing and nesting;
- to explore real data sets using a variety of graphs and numerical methods;
- to assess how well the standard assumptions of analysis of variance (ANOVA) fit a data set, and if the fit is poor, to choose a suitable remedy such as transforming to a new scale;
- to decompose any balanced data set into "overlays" (components corresponding to the factors in the design) and to find the parallel decompositions of the sums of squares and degrees of freedom;
- to construct the interval estimates and F-tests of formal inference; and
- to interpret numerical patterns and formal inferences in relation to the relevant applied context.

Approach

Four main features characterize the approach of the course: modest technical demands, a focus on a small set of recurring basic ideas, an emphasis on examples and context rather than derivations, and a preference for active learning.

- *Modest technical demands.* To make the material accessible to students who do not have strong algebraic skills, the course presents ANOVA visually, without formulas, by representing factors of a design as partitions of the data, and computing the usual summaries by "sweeping out" the averages that correspond to the partitions. (This approach is an elementary analog of taking orthogonal projections into subspaces of Rn.)
- *Small set of recurring ideas.* Our aim is for students to develop a flexible ability to apply a few broad principles. One instance is the approach to decomposition of the data via partitions and averages, described above. Another is an approach to design based on two principles (randomization and blocking) for assigning treatments to experimental units, and one principle (factorial crossing) for choosing a treatment structure.
- *Examples and context.* Students begin by learning, via a series of concrete examples, four specific designs (completely random, complete block, Latin square, split plot/repeated measures) built from the three basic principles. Later, these same principles are used to extend each of the designs to a family of similar designs. Over the course of the semester, specific data sets return again and again, so that students get to know them and rely on them to learn the general principles and structures.

- *Active learning.* Applied statistics lends itself naturally to discussion: for example, on the merits of completely random versus complete block designs to see whether carpeting raises levels of airborne bacteria in hospital rooms, on the choice between logarithms and reciprocal roots for analyzing the effect of day length on the concentration of a neurotransmitter in the brain of a hibernator, or on the way a few mildly deviant observations should qualify conclusions from the F-test in a study of babies learning to walk. Discussions of these and other issues arise easily when the course emphasizes real data sets and their applied context. For the same reason, writing assignments are easy to incorporate into the homework. Both the discussions and the writing help prepare the way for a substantial term project that students complete over the second half of the semester, and present first orally to the rest of the class and then in a final paper.

APPENDIX C: Archaeometrics — An Interdisciplinary Introduction to Statistics

(Don Bentley, Department of Mathematics, Pomona College)

Archaeometrics, an interdisciplinary course, was first offered at Pomona College during the Spring semester of 1996–97. As the name indicates, the topics addressed in the course are applications of statistical models to the field of archaeology. There are no prerequisites for the course, and it satisfies the general education requirement of a course in statistical reasoning. It was designed to be attractive to students in the humanities who feel less than secure in their quantitative skills and need a course to meet this General Education requirement.

Archaeology is an inductive science. Information gathered in excavations is used to make inference about the history and culture of the societies that occupied the studied locations. The discipline of statistics provides tools to assist in inductive reasoning. Dever, in writing on the "New Archaeology"[1] points out that "… the *general* influence of the explicitly scientific school is seen in the deliberate development of research design, in the emphasis on problem solving, and in the testing of hypotheses in general, which increasingly characterized the more sophisticated American project of Syro-Palestinian archaeology …." The structure he describes falls right in the center of the field of statistical inference.

While attending a course titled Archaeology of the Lands of the Bible, this author was continuously amazed at the number of instances in the assigned readings where statistical modeling could have been applied to assist the archaeologists in making inferences from their data. In some instances the models would have suggested further questions to ask of the data, or provided direction for seeking additional data. In other instances, the inferences made were logically invalid. By the end of the semester he was convinced that there was a great need for the new field, archaeometrics, and it would provide a great opportunity for an exciting interdisciplinary course.

In considering the structure of the course it was decided that it should be directed towards students with an interest in areas in which archaeology is an important tool, such as religion, anthropology, and history. To attract this audience precludes setting any mathematics or statistics prerequisite. But the course should also appeal to students in sciences who are seeking areas of application for their concentrations. For this reason there is no prerequisite for the course.

The statistics topics presented in the course are different from those found in the traditional introductory statistics course. They emphasize those models and methodologies that are appropriate for the data and problems encountered in archaeological investigations. As an example, the Poisson distribution is extremely important for archaeological studies as it is, in many instances, the appropriate model to describe the behavior of random data. It is quite common that only a small percentage of a tel is excavated (on the order of 5%). Thus excavations are restricted to extremely localized areas which causes a great deal of dependence to exist within the data. Students must gain a good understanding of the consequences of this dependence. Sampling bias is also very common in archaeological studies since the decision as to where to search for data is made by determining the most likely areas to produce the type of information desired. Another extremely common problem is that inferences are frequently made on the basis of sparse data. Therefore, the concept of power of a test needs to be emphasized.

Above are examples of topics in statistics that are important for the archaeologist to understand when making inferences from collected data. Another area of statistics that is not covered in introductory statistics courses, but could be of benefit to biblical archaeologists, is logistic regression. This methodology would assist archaeologists in determining which specific potential excavation site is most likely to yield particular types of artifacts.

---

[1]William Dever, "Syro-Palestinian and Biblical Archaeology." The Hebrew and its Modern Interpreters, eds. Knight and Tucker, (Chico, CA: Scholars Press, 1985),49.

APPENDIX D: A Data-Oriented, Active Learning, Post-Calculus Introduction to Statistical Concepts, Methods, and Theory

(Allan Rossman, Department of Mathematics, Dickinson College; Beth Chance, Department of Statistics; Cal Poly-San Luis Obispo; Karla Ballman, Mayo Clinic)

We have each taught and are continuing to develop a course that provides post-calculus students with an alternative introduction to probability and statistics that focuses on data and active learning. The course incorporates many of the features that have emerged in "Stat 101" courses (analysis of genuine data, focus on conceptual understanding, student exploration and hands-on activities, use of computer tools) and provides a more balanced introduction to the discipline of statistics than the standard sequence in probability and mathematical statistics. The course differs from the algebra-based introductory course by utilizing students' calculus knowledge and mathematical abilities to explore some of the mathematical framework underlying statistical concepts and methods. We have taught this course to mathematics, statistics, computer science, economics, science, and engineering majors. Previously, we utilized a text that focused on data analysis concepts, such as Moore and McCabe, while developing supplementary activities to encourage student explorations of the mathematical underpinnings. We are now incorporating these activities into a stand-alone text.

Course Principles and Goals:

The guiding principles of the course are for students to:

- Develop understanding by conducting investigations of statistical concepts and properties.
- Learn about probability models in the context of statistical ideas, applied to real data.
- Utilize their mathematical skills such as their knowledge of functions, graphically and analytically, as well as counting techniques and calculus optimization methods.
- Develop an assortment of problem-solving skills as they approach data from graphical, numerical, and analytical perspectives.
- Use technology as a tool for simulations, graphical displays, exploration, etc.
- Analyze data from a variety of fields of application, including data from scientific studies, popular media, and self-collected.

By the end of the course, we hope students will have:

- acquired proficiency with a variety of specific data-analytic techniques, including exploratory data analysis, confidence intervals, tests of significance, t-tests and intervals, regression analysis, contingency table analysis, analysis of variance;
- gained an understanding, at both conceptual and mathematical levels, of fundamental statistical ideas, such as variability, randomness, distribution, association, transformation, resistance, sampling, experimentation, confidence, significance, power, model.

We also aim to help students develop skills of:

- reading technical material
- using computer software
- working productively with peers
- expressing themselves in written and oral presentations

Content:

Students are introduced to statistical ideas at the conceptual level in a genuine context, then to methods of analysis, and finally to the theoretical foundation. Students then continually revisit important ideas in new settings. For example, an early unit on randomization and significance begins with the results from a scientific study to motivate the question, uses simulation to explore the likelihood of obtaining such sample results by randomization alone, and then develops the combinatorial tools necessary to determine the exact p-value. The focus throughout this unit is on the interpretation of p-value and the proper conclusions that can be drawn from an experimental study. The next unit then compares these results to the analysis and conclusions that can be drawn from an observational study. A tentative outline of the course syllabus includes:

- Variation, randomization, comparisons, significance (analysis of two-way tables)
- Observational studies, confounding, association vs. causation
- Sampling, binomial model, tests of significance, power
- Large-sample approximations, normal distribution
- Confidence intervals for categorical variables
- Analysis of quantitative variables (including t-tests and intervals)
- Analysis of bivariate data (regression analysis, analysis of variance)

Use of Technology:

The course is taught in a computer-equipped classroom and makes fairly extensive use of technology. The computers play an important role in this course in three ways:

- performing calculations and creating graphics necessary for analyzing data
- conducting simulations to approximate long-run behavior of random phenomena
- addressing "what if" questions, allowing students to explore probability and& statistics concepts

A statistical software package is essential for teaching the course; we use Minitab. We also use the spreadsheet package Excel for facilitating students' investigations. Java applets created specifically to help students to visualize and explore statistical concepts are also used.

Assessment:

Student work in this course is evaluated on the basis of regular assignments, quizzes, fairly standard exams, and more extensive written assignments and projects. The homework assignments are a combination of interpretation and presentation of results, practice on techniques, and mathematical extensions. Students are also expected to be able to utilize a statistical package and interpret its output, and conduct simulations.

More information about this course and project can be found at `www.rossmanchance.com/scmt/`.

APPENDIX E: A Case-Studies Based Mathematical Statistics Course

(Deb Nolan, Department of Statistics, University of California at Berkeley)

Nolan and Speed have developed a course and accompanying text (Stat Labs) that teaches undergraduate upper-level mathematical statistics through the use of in-depth case studies. They call these case-studies "labs."

It is via the labs that mathematical statistics is introduced, which leads to an integration of statistical theory and practice in a way not commonly encountered in an undergraduate course. The labs raise scientific questions that are interesting in their own right, and they contain datasets for use in addressing these questions. The context of the scientific question is the starting point for developing statistical theory.

The labs are central to the course. Each lab is a substantial exercise with nontrivial solutions that leave room for different analyses. They serve to motivate and to provide a framework for studying theoretical statistics, and they give students experience with how statistics can be used to answer scientific questions. An important goal of this approach is to encourage and develop statistical thinking while imparting knowledge in mathematical statistics. Through the labs students develop their quantitative reasoning and problem-solving skills in a broad multidisciplinary setting. They become practiced in communicating their ideas orally and in writing, and they become versed in the use of statistical software.

Audience and Course Prerequisites

The labs are designed for use in a mathematical statistics course for juniors and seniors. They have been used by Nolan and Speed for two such courses, one for mathematics and statistics majors, and one for engineering and computer science majors. Both courses require students to have studied calculus for two years and probability for one semester. There is no statistics prerequisite for the course. The course enrollments range from 20 to 60 students. Classes meet three hours a week with the instructor and one hour a week with a teaching assistant.

In a typical semester, roughly one week is spent on summary statistics, and about three weeks are spent on each of the following four areas: sampling, estimation and testing, regression and simple linear least squares, and analysis of variance and multiple linear least squares.

Lab Organization

The material for a lab is divided into five main parts: an introduction, data description, background material, investigations, and theory. In the introduction, a clear scientific question is stated, and motivation for answering it is given. The question is presented in the context of the scientific problem, and not as a request to perform a particular statistical method. Documentation for the data collected to address the question is provided next. It includes a detailed description of the study protocol, as appropriate. Material to put the problem in context is provided in the background section. Information is gathered from a variety of sources, and is presented in non-technical language. The idea is to present a picture of the field of interest that is accessible to a broad college audience.

Suggestions for answering the question posed in the introduction are provided next in the investigations section. These suggestions are written in the context of the problem, using very little statistical terminology. The ideas behind the suggestions vary in difficulty, and are grouped to enable the assignment of subsets of investigations. Also included are suggestions on how to write up the results, e.g., as an article for a widely read magazine; as a memo to the head of a research group; or as a pamphlet for consumers. Often included in the report will be an appendix containing more technical material.

The theoretical development of the statistical concepts and methodology appear after the problem is introduced, at the end of the lab handout. The material includes information on general topics in statistics, as well as topics more specific to the individual lab.

Student Work

The labs are chosen by the instructor according to the topic (theoretical or practical) and the background of the students. They are divided between those labs that are discussed primarily in lecture and those that require students to do extensive analyses outside of class and to write short papers containing their observations and solutions. Typically students write reports for four labs, about one for each of the five main topics in the course. The students find this work very challenging, and they typically work in groups of twoor three on their lab assignments.

Bringing the computer into the theoretical course enables us to go far beyond the traditionally small, artificial examples found in textbooks. But care is taken to keep the demands made upon students at an appropriate level. Assistance on how to use statistical software is provided in the weekly section meeting, and handouts with sample code are also provided. Often the section meets in a laboratory room, where students double up at workstations to work on the assignment and the teaching assistant provides advice as needed.

Nolan and Speed have found that when using these labs several fundamental changes take place in the classroom. The format of lectures changed. More time is spent on determining how to answer general scientific questions using statistical analyses and on deriving a statistical method from its application to a specific problem. Less time is spent covering many small examples constructed to illustrate a single statistical technique, yet all the basic material traditionally covered in the course is still covered.

Lectures also include: discussion of the background to a particular problem, where students who have taken courses in related fields can bring their own expertise to the discussion; and motivation of the theoretical material through discussion of how to address a problem from a lab. In addition, for labs on which students are to write reports, we hold regular question and answer periods where students raise concerns about their work. Roughly about one class period in three is spent on these types of activities. The remainder of time is spent in a more traditional presentation of theoretical results.

More information about Stat Labs can be found at `www.stat.berkeley.edu/users/statlabs`.

APPENDIX F: Time Series Analysis—An Alternative Introduction to Applied Statistics for Mathematics Students

(Robin H. Lock, Mathematics Department, St. Lawrence University)

Time Series Analysis is offered at St. Lawrence University every other year in the spring semester. The course is dual listed under mathematics and economics so students may use it as an upper level elective towards either major. The primary audience is juniors and seniors with an occasional sophomore and mostly mathematics, economics, or combined math/econ majors and statistics minors. Prerequisites are just two semesters of calculus. Although students may have already seen some statistics through our introductory service course, mathematical statistics, or an econometrics course, we assume no particular background in statistics. Thus the course serves as a first exposure to statistical ideas for some mathematics students and, fortunately, the content is sufficiently disjoint from other offerings that students with previous experience will see lots of new material. This flexibility increases the pool of students for the course.

A course in time series analysis offers a number of unique opportunities for introducing mathematically oriented students to the applications of statistics. The data structures encountered in the study of time series are typically very straightforward—often just a single univariate series of historical values. Yet the statistical methodology used in the analysis can require a broad range of mathematical sophistication while still illuminating fundamental concepts of applied statistics. Real world applications are immediate and compelling. Wouldn't many students today be interested in methods for predicting where the stock market will go?

The general notions of an underlying model for some real world phenomenon, estimation of its parameters from data, and diagnostic checking of the model assumptions are central themes in statistics. The models encountered in forecasting are fairly intuitive, yet can be used to effectively illustrate important statistical principles such as parsimony, variability in parameter estimates, construction of prediction intervals, and criteria for choosing between competing models. The clear interaction among the "identify," "estimate," and "forecast" steps of the Box-Jenkins approach can be fruitfully applied to many other statistical situations. The analysis of residuals to check model assumptions, suggest alternative models, or gauge the accuracy of the fit is a featured part of time series methodology that is often neglected in traditional introductions to statistics. Similarly, statistical graphics—plots such as the time series itself, a differenced or transformed series, residuals, or sample autocorrelations—are used at many points throughout a time series analysis. This integration of graphics into all phases of an analysis is an important part of modern statistics.

What about mathematical content? A key point for attracting mathematically oriented students to such a course is the inclusion of nontrivial applications of mathematics. These must go beyond simply "plugging & chugging" with more complicated formulas than the standard introductory service course. As one example, consider the duality between autoregressive (AR) and moving average (MA) models. The question of stationarity/invertibility of such models, depending on the location in the complex plane of the roots of a characteristic polynomial determined by model parameters, can be quite challenging. In fact, the process of inverting an MA model to get an AR model (of infinite degree) may seem to be only of mathematical interest, until one sees an example where a complicated AR model requiring several parameters is effectively replaced by a more efficient MA model with perhaps just one parameter. The use of the difference operator on a time series provides an interesting analogy to the differentiation process that is familiar to most students from calculus. In fact, the use of operator notation (differencing and backward shifts) in specifying and manipulating many time series models is good experience for mathematics students in dealing with the general notion of an abstract operator. Of course, one can also see the traditional calculus-based derivation of least squares estimates in regression, with a neat special case for students to work on their own when the dependent variable can be assumed to take on the values $t = 1, 2, \ldots, n$. One could even dive into the spectral analysis of a time series, although that would probably be beyond the mathematical sophistication of most undergraduates.

Our time series syllabus typically includes the following topics:

1. **Basics**: models, parameters, estimators, residuals, mean squared errors, probability, normality, expectations, autocorrelation.
2. **Linear regression on t**: least squares estimation, tests for parameters, transformations, diagnostics on residuals, confidence intervals for forecasts.
3. **Exponential smoothing**: simple and double exponential smoothing, choice and effect of smoothing constants, confidence intervals.
4. **Box-Jenkins methodology**: development of the general ARIMA model with emphasis on special cases, patterns in autocorrelations and partial autocorrelations, invertibility, stationarity, differencing, autoregressive / moving average duality, model identification, estimation, and forecasting (with considerable computer assistance).
5. **Seasonal models**: based on cosine trends, ANOVA means, or seasonal ARIMA terms.
6. **Models based on one or more other series**: simple and multiple regression, selection of "predictor" series.
7. **Mixed models**: combinations from among (2), (4), (5), and (6) above.

Some additional topics that we typically do not get to (either due to time constraints or complexity of the material) include Winter's method, formal estimation methods for ARIMA models, matrix notation for regression, Fourier methods and spectral analysis.

The course relies heavily on actual data for motivation and illustration. Fortunately, a wide variety of excellent sources (including websites) for time series data are readily available. In contrast to some other areas of statistical application, published reports involving time series seem to be more likely to include all the data, either explicitly or graphically, rather than just summary statistics. Thus students have relatively little difficulty finding interesting data on their own for projects. We use Minitab as a statistical package for computation and graphics, although a number of specialized time series packages would also work well.

In summary, a course in time series analysis can be used as a mechanism for attracting and introducing mathematics students to the field of applied statistics. They can encounter many of the fundamental principles of statistical thinking in a mathematically interesting setting and, hopefully, be encouraged to pursue further work in statistics.

APPENDIX G: A First Course in Statistics for Mathematics Majors at Duke University

(Dalene Stangl, Institute of Statistics and Decision Sciences, Duke University)

At Duke University, the introductory statistics course for mathematics majors was designed with the belief that real-world applications motivate and drive the field of statistics. With this in mind two faculty, Mike West and Robert Wolpert, have designed a course that balances theory and application. Student curiosity is piqued on day one, when they are presented with two quotes:

Bruno de Finetti, on foundations: *Probability does not exist.*

George E.P. Box, on mathematical and statistical modeling: *All models are wrong, but some are useful.*

The course is an introduction to the concepts, theories and methods of statistical modeling and inference. It focuses primarily on ideas and methods of modern Bayesian statistical science, but it also teaches and contrasts classical methods. Statistics is a vast field, and a first one-semester course must be a broad introduction overlaying key conceptual ideas. The over-arching goal of the course is to provide this introduction, exploring the foundations of scientific reasoning and inference and rousing curiosity via applications in medicine, genetics, policy, astronomy, physics, economics, finance, education, and many other fields. Students use statistical software to explore data and statistical models. Students learn the theory and software that enable scientific learning in the face of uncertainty.

Topics covered:

- Reasoning with Probabilities: Scientific Learning and Inference
- Learning from data and observation: Pre-data and Post-data probabilities
- Bernoulli trials and binomial sampling models
- Likelihood and maximum likelihood
- Exponential, Poisson, Normal and other sampling models
- Simulation of probability distributions for inference and model assessment
- Sampling distributions and pre-data probabilities
- Perspectives on statistical inference: Classical and Modern
- Significance testing and p-values
- Honest prediction
- Linear regression models, including various special classes
  (factor models, multiple regression, time series regression, ...)
- Graphical data display and exploration
- Statistical computing with *S-Plus*

The course draws on material from the traditional text *Probability and Statistics* by Morris H. DeGroot, published by Addison Wesley (2nd Edition). This book covers basic elements of both Bayesian and non-Bayesian approaches to statistics and has been a standard introductory text for many years. Many other texts cover similar material on the non-Bayesian side. Chapters 1–5 provide in-depth coverage of prerequisite probability theory. Supplementary material beyond the scope of the text, particularly on applied statistical methods is provided in handouts from the website `www.stat.duke.edu/courses/` under STA114. Supplementary texts include *Bayesian Data Analysis* by Andrew Gelman, John B. Carlin, Hal S. Stern and Don B. Rubin, published by Chapman & Hall. This book goes beyond the scope of this course, but it is a truly excellent text for both statistical modeling and applications. More extensive material on the theory side, but less so on the application side, appears in *Bayesian Statistics: An Introduction* by Peter M Lee, published by Arnold UK and distributed in the USA by Wiley (2nd Edition). The software used for the course is *S-Plus*. This software is available in student edition for Windows from Duxbury Press.

Teacher Preparation: K–12 Mathematics

CRAFTY Curriculum Foundations Project
Michigan State University, Division of Science and Mathematics Education (DSME) November 1–3, 2000

Sharon L. Senk, Brian Keller, and Joan Ferrini-Mundy, Report Editors
Sharon L. Senk, Workshop Organizer

Summary

As noted in the Mathematical Education of Teachers (CBMS, 2001, p. 7), "teachers need to understand the fundamental principles that underlie school mathematics, so that they can teach it to diverse groups of students as a coherent, reasoned activity and communicate an appreciation of the elegance and power of the subject." Although the mathematical needs of teachers depend upon the grade level at which they intend to teach, substantial knowledge of content is needed by all teachers. For instance, elementary school teachers need a solid understanding of place value and the distributive property to help their students make sense of operations with whole numbers and decimals. High school mathematics teachers need to understand that same content to help students make sense of polynomial arithmetic.

Below are five principles concerning the mathematical preparation of teachers of all grade levels that emerged during this CRAFTY Curriculum Foundations Workshop. Though not officially adopted at the workshop, these principles reflect the workshop discussions and the views expressed by most of the workshop participants.

1. Mathematics courses for future teachers should develop "deep understanding" of mathematics, particularly of the mathematics taught in schools at their chosen grade level.
2. Tools for teaching and learning, such as calculators, computers, and physical objects, including manipulatives commonly found in schools, should be available for problem solving in mathematics courses taken by prospective teachers.
3 Mathematics courses for future teachers should provide opportunities for students to learn mathematics using a variety of instructional methods, including many we would like them to use in their teaching.
4. Faculty involved in the preparation of teachers of mathematics should engage in study and discussion of how people learn mathematics.
5. Greater communication and cooperation is necessary among all stakeholders in the mathematics preparation of teachers.

The mathematical preparation of future teachers is a complex and often controversial enterprise. Given this, as well as the diverse nature of the workshop participants, it should come as no surprise that consensus was not always achieved in our discussions. However, the following narrative represents the editors' best attempt to summarize the dominant opinions expressed during the workshop sessions and provide further details in support of the five principles stated above.

Narrative

Introduction and Background

According to data collected by the National Center for Education Statistics, mathematics teacher preparation is the second largest of the specialty areas considered by the Curriculum Foundations Project. Only business has more undergraduate majors.

Mathematics teacher preparation is not only a large endeavor it is also quite complex. Teacher preparation is influenced by certification requirements set by the states and recommendations provided by several professional societies. In some states elementary teachers are not required to take any courses about the mathematics they will teach in schools. In others, they are required to take several courses that focus on deepening their understanding of number systems and geometry. In addition, teacher education programs in universities often participate in state or national accreditation processes that also impose specific standards for student learning.

Regardless of certification regulations or accreditation expectations, the mathematical requirements of future teachers vary considerably with the level at which the student intends to teach. Elementary teachers are usually prepared to be generalists who take courses in many departments. Consequently, they are required to take relatively few courses in mathematics. In contrast, teachers of high school mathematics are expected to be specialists, and often must complete a full mathematics major. In states where special certification for mathematics teachers of middle grades exists, the mathematical requirements for such teachers are generally between those of future elementary and high school teachers.

Consistent with the charge from CRAFTY, most participants in this workshop represented the discipline of mathematics education. Invited participants included classroom teachers with elementary, middle and high school teaching experience, college professors who teach methods courses and supervise field placements of prospective teachers, and college professors who teach courses in the mathematical content needed by teachers. Among the latter were both mathematicians and mathematics educators. In addition, professors and graduate students from the departments of mathematics, teacher education, and educational psychology at Michigan State University also attended parts of the workshop.

In order to give guidance to colleges and universities, professional societies have issued recommendations for the mathematical preparation of future teachers. Two of the most influential were *A Call for Change* (MAA, 1991) and *Professional Standards for Teaching Mathematics* (NCTM, 1992). Most recently, the CBMS member societies collaborated to produce a report called the *Mathematical Education of Teachers* [MET] (CBMS, 2001). The MET report recognizes the mathematical needs of teachers preparing to work at the elementary, middle grades, and secondary school levels.

A revised draft of the MET report was available on the web in September 2000. Because the recommendations in that draft overlap with the issues identified by the CRAFTY Curriculum Foundations Project, the planning committee for this CF workshop decided to use the MET report as a basis for discussion at the workshop, and chose to interweave the themes provided by the CF project within that framework. The workshop opened with a plenary address about the MET report. During the next day and a half workshop participants heard a series of presentations on specific issues related to the preparation of teachers at elementary, middle, and high school levels. Speakers at these plenary sessions included both mathematicians and mathematics educators. Following each presentation participants met in breakout groups to discuss issues in the preparation of teachers at that level. Our goal was for participants to discuss the MET report, and to provide reactions and perspectives based on these discussions to CRAFTY regarding the specific mathematical needs of future school mathematics teachers during their first two years in college.

What constitutes the first two years of college mathematics for a future teacher varies widely in institutions across the country. Future high school teachers usually take calculus during their first two years, while future elementary teachers seldom do. When mathematics courses about number or geometry are offered for future elementary teachers, in some universities they are taken during the first two years, but in others they are taken later. In our discussions, the conference participants leaned intentionally on the side of breadth.

Participants were assigned to breakout groups to ensure that various levels of teacher preparation and different academic departments were represented. On the last morning of the workshop, following a presentation about the role of community colleges in the preparation of teachers, participants chose one of elementary, middle or high school teacher preparation and developed initial recommendations to CRAFTY about mathematics during the first two years of college for that level. Recommendations from each of these final breakout groups were presented to the whole group. Because of this structure no one at the workshop heard all the discussions taking place. Because of the size and complexity of the teacher preparation enterprise in the United States and the diversity of the workshop's participants, many perspectives on issues were presented in the breakout sessions and spirited debates often ensued. Although there was some time for whole group discussion at the end of the workshop, there was no time to even begin to develop consensus; rather, we were able to raise the issues as they emerged from the subgroups.

What follows is drawn from the presentations and discussions that took place at the workshop. Space does not permit us to report all of the comments and examples made at the conference. The people whose names are identified were among those who spoke at the plenary sessions. Remarks set off in italics indicate ideas presented to the whole group and recorded in papers provided to the organizers after the workshop. The five principles noted in the summary and in bold on the following pages reflect syntheses by the authors of this report of ideas that seemed to be accepted by most of the workshop participants.

One caveat: The report that follows contains many suggestions about what conference participants thought might be added to courses that are either part of the general education requirements for all college students or for specific courses for future teachers. Given the time available at the workshop participants did not discuss what they might delete from the college curriculum, or how it might be reorganized, to make room for these newer areas of emphasis. Nor did participants discuss how they might address situations in courses, e.g. calculus, in which needs of future teachers might be somewhat different than needs of future engineers or others.

Understanding and Content

Mathematics courses should enable future teachers to develop "deep understanding" of mathematics, particularly of the mathematics taught in schools at their chosen grade level.

"Deep understanding" is a term used without definition in the MET report. Workshop participants discussed the ambiguity of the term "deep understanding" and did not come to resolution about an exact meaning. Participants' usage seemed to include conceptual understanding that goes beyond procedural fluency, including being able to represent concepts in multiple ways, explain why procedures work or recognize how two ideas are related. It also involves being able to solve problems and to make connections among mathematical topics or between mathematics and other disciplines.

All students benefit from studying connections within a topic, e.g., how graphs, tables, words, and symbols may all represent the same function, and between topics e.g., how algebra and geometry are related. In addition, prospective teachers benefit from knowing about the connections between the mathematics they are studying in college and the mathematics they will have to teach. Thus, deep understanding should be a goal of all mathematics courses taken by future teachers, whether they are courses satisfying general requirements for graduation or specialized courses for teachers.

Deborah Ball and Hyman Bass argued in their presentation that looking at the actual practice of teaching provides insights into what mathematical understanding is needed by future elementary teachers. Videos of a 3rd grade class taught by Ball suggest that a teacher with a deep understanding of number theory and the role of definitions, conjectures and proofs is able to engage young children in mathematical study that emphasizes meaning. This suggests that mathematics courses taken by prospective teachers should develop understanding of both mathematical content and mathematical processes such as defining, conjecturing and proving.

Courses in algebra or college algebra often satisfy general education requirements and serve as prerequisites for specialized content courses for many elementary and secondary teachers. The content and methodologies of these pre-calculus courses should be examined by mathematics departments and committees of the MAA such as CRAFTY and the CUPM. Algebra courses should be redesigned to reflect the deep understanding recommended in the MET report. In general, there should be less symbolic manipulation, and more modeling and problem solving from both a discrete and continuous perspective. Emphasizing connections between algebra, arithmetic, and geometry is one way to develop understanding in these courses. An example of connections that could be explored involves examining the relation between the FOIL method of multiplying binomials, strategies used for mental multiplication, and areas of rectangles. Such revised algebra courses may also better serve students in other majors.

Calculus often serves as the entrance requirement into a major in mathematics. The desire for fluency in finding derivatives and integrals symbolically should be balanced with questions asking about the implications of the mathematics being studied. Problems requiring students to explain or justify their thinking should become more prominent in calculus courses. Curt Bennett, a conference participant, suggested that having students discuss issues such as the potential equality of .9999... = 1.0 or how round-off error affects calculations might deepen their understanding of both topics in the school curriculum and topics in calculus.

Bennett also suggested that calculus courses spend more time on infinite sequences and series. He claimed that at present

> *students learn a plethora of techniques for proving whether or not a series converges, but they gain little understanding of why the techniques work, or even why you might want to know them. Spending more time on this topic would allow students to explore various important series, and connect them more directly to the idea of number. In particular, infinite decimals could be explored in more depth beyond repeating decimals, something all our students would benefit from. For example, students could explore the series.*
>
> $$\sum_{n=1}^{\infty} \frac{1}{n!}.$$
>
> *Similarly, they could explore other interesting series that converge to well known numbers preparing them for the study of Taylor's series at the same time.*

To be able to spend more time on sequences and series, Bennett would spend less time on integration techniques. Several workshop participants mentioned that some pre-calculus and calculus courses seem like collections of miscellaneous techniques to students. They suggested that each mathematics course should have a theme. Mike Lehman, a secondary school teacher, said that every mathematics course "should be about *something*." Others suggested that every course should emphasize a few *big ideas*, and that the teacher should explicitly refer to the development of those ideas during the year. Some of the big ideas mentioned were: equivalence, function, transformation, composition, representation, proof and dimension. Michaele Chappell noted weaknesses in spatial visualization among her students and called for increased attention to 3-dimensional geometry during the first two years of college. The book *On the Shoulders of Giants* (Steen, 1990) was suggested as a good source for ideas about big ideas in geometry and other fields.

Conference participants generally concurred with the recommendations in the MET report that emphasized the need for future teachers to develop a deep understanding of the mathematics they will need to teach in schools. Thus, all mathematics teachers from elementary through high school should develop fundamental understanding in four areas: number and operations, algebra and functions, geometry and measurement, and data analysis, statistics and probability. Detailed lists of important topics in each area are given in the MET report. Chapell pointed out that many future teachers of secondary school mathematics taking her methods courses have not studied statistics or probability. Ideally, experiences in these subjects should be part of the first two years of college mathematics for future teachers.

Concern was expressed that the expectations expressed in the MET report for the mathematical preparation of elementary teachers were not strong enough. However, there was general agreement that the mathematics courses taken by elementary teachers, particularly those emphasizing number and geometry, should be completed during the first two years of college and before the student takes courses in the methods of teaching and that the mathematics courses should be coordinated with the methods courses.

Many workshop participants supported the recommendation in the MET report that mathematics in the middle grades (grades 5–9) should be taught by mathematics specialists. Many felt that programs should also be developed for elementary teachers who want to become "mathematics specialists." To develop programs for elementary or middle grades specialists some existing courses would probably have to be redesigned and new courses might have to be created.

Ira Papick described a course being developed at the University of Missouri in algebraic structures for middle grades teachers. In this course Papick introduces students to fundamental notions in number theory (e.g., greatest common divisor, least common multiple, the Euclidean algorithm) through problems taken from NSF-funded middle school curricula. Papick claims that

> *Courses of study relating core middle school algebra curricula to important applied and theoretical topics of university algebra ... empower teachers with new algebraic tools and perspectives, which in turn better prepares them to prompt important questions and enables them to convey a wide spectrum of algebraic ideas. Such courses would provide middle grade mathematics teachers with a strong mathematical foundation and directly connect the mathematics they are learning with the mathematics they will be teaching.*

Tony Peressini and several colleagues have developed a new course that focuses on connections between high school and college mathematics as well as connections within high school mathematics (Usiskin, Stanley Peressini, and Marchisotto, in press). This advanced perspective is also mindful of the historical and conceptual evolution of mathematical theory and school mathematics. For instance, the discussion of the real number system in this course

> *begins by presenting the real number system as it is typically presented in high school—as the set of points on a number line or as the set of decimal numbers. This includes a careful discussion of the decimal representation of the real numbers based on the nested interval property of the number line. Then the algebraic structure of the real number system is placed in the context of algebraic structures, first as a field, then as an ordered field, and finally as a complete ordered field—the structure that is often used to define the real number system in college courses. Finally, the connection returns to the high school setting of the real number system by describing how a decimal representation of each element of a complete ordered field can be constructed.*

The current calculus sequence and other requirements for a mathematics major often result in the study of the mathematics for teaching, such as the courses developed by Papick or Usiskin et al., being taken during the last two years of the undergraduate experience. This may not be enough time to help future teachers develop deep understanding of school mathematics. This concern might be addressed with new models of the mathematics major that allow future teachers to engage in connecting college mathematics to school mathematics earlier in their studies. One option is to design calculus courses for prospective teachers that are set up to make explicit connections to school mathematics. Another option is to offer other courses, say discrete mathematics or geometry, before calculus, where such connections can be made more easily.

Technology

Tools for teaching, such as calculators, computers, and physical objects, including manipulatives commonly found in schools, should be available for problem solving in mathematics course taken by prospective teachers.

Conference participants did not have a thorough conversation about technology, and there were a variety of views about the appropriateness of its use in the school curriculum. However, some examples were discussed. Because technology can generate data so quickly, it can be a useful tool for making conjectures. However, in order to develop understanding, the use of technology should be accompanied by analytic methods or classroom discussion. For instance, in a course in which future elementary teachers study decimals one might use a calculator to determine the sum $.\overline{7}+.\overline{8}$. Then students could be asked to find the sum using the fraction equivalents of the decimals and compare the results.

When students begin studying the trigonometric functions and limits, having them explore the limit of $\sin(x)/x$ on a calculator can lead to interesting results. Bennett reported that when he did this in his classes, about half of the students came up with the answer 1, while the other half came up with the answer .017. Other participants reported that they had experienced similar results. Analytic techniques verify that the limit of $\sin(x)/x$ equals 1. Thus, students naturally question why calculators give two seemingly different answers. This disparity presents the opportunity to bring up the question of what units are being used, and students have the opportunity to discover, for themselves, a reason to use radian measure for angles.

Spreadsheets and the graphing and table-generating features of graphing calculators can be effectively used to solve problems about functions and families of functions. At present, the study of functions, including the use of exponential functions to describe both population growth and compound interest, is found in many middle and high school curricula. Marjorie Economopoulos and Marian Fox reported that faculty at of Kennesaw State University (KSU) have developed materials using spreadsheets, graphing calculators, and existing school mathematics curricula to help pre-service teachers deepen their understanding of functions. Mathematics courses at KSU also give prospective teachers opportunities to analyze and model data taken from secondary school curricula using the statistical features on a graphing calculator or a statistics package on a computer. Participants also suggested that calculus and other courses for future teachers make use of computer algebra systems to investigate equivalence of expressions and transformations of functions.

Future teachers should be able to use tools such as tiles, cubes, spheres, rulers, compasses, and protractors to deepen their understanding of the mathematics they will have to teach, including 2-D and 3-D geometry and measurement. The use of electronic drawing tools such as the *Geometer's Sketchpad* and *Cabri* is also recommended for use in geometry courses for both elementary and secondary teachers. The dynamic capabilities of these tools allow both students and teachers to test conjectures relatively easily.

Instructional Techniques

Mathematics courses for future teachers should provide opportunities for students to learn mathematics using a variety of instructional methods, including many we would like them to use in their own teaching.

This perspective arose from the belief that instruction in college mathematics classes should involve more than lecture. Instructors should include various techniques for engaging students actively in solving problems. This could include, whenever appropriate, having students solve problems or discuss strategies with a partner or small group, and engaging the whole class in discussion. Instructors and students should be encouraged to solve problems in more than one way, to explain their reasoning, and to describe how the mathematics they are doing today is related to mathematics done earlier. The use of such techniques in undergraduate mathematics courses could help develop the college students' understanding of the mathematical content, and provide an instructional model for the future teachers to emulate.

Proof and justification are an integral part of mathematics in comparison to other sciences. There was virtually universal agreement among workshop participants that reasoning and proof should be a theme in all college mathematics courses, beginning at least with calculus. This is not a call for reintroduction of delta-epsilon proofs into calculus or for an emphasis on axiomatics and formality. Rather, in the introductory courses there should be an emphasis on developing basic mathematical reasoning and communication skills by asking students to explain their thinking or to justify their responses based on definitions. Specific proof techniques need not be taught until later, and at that time students should be given many opportunities to create their own proofs in an effort to promote their own understanding of mathematics. Most elementary mathematics texts do not emphasize reasoning and proof and as a result new materials—in the form of activities, problem sets or entire books—may need to be developed.

Prospective teachers need the opportunity to work on extended problems, perhaps through projects, as well as on short assignments. The content and methods of assessment should be consistent with the content and methods of instruction. Tests, quizzes and projects should measure the students' understanding of the topics emphasized in the course. Questions should ask about representations, connections, reasoning, and problem solving, as well as computations and algorithmic procedures.

Teachers of mathematics courses for future teachers should be familiar with the mathematics content and expectations of elementary schools, and with elementary school children. Mathematics courses designed primarily for future teachers should not be assigned to inexperienced Teaching Assistants unless substantial training and on-going support are provided. More generally, in order to provide high quality mathematics instruction for future K–12 teachers, there needs to be support for the professional development of their college and university teachers. In particular, teaching assistants and instructors should work as apprentices under the supervision of experienced teachers.

Conference participants were generally supportive of the idea that prospective teachers should experience mathematics instruction that engages them in actively doing mathematics on a regular basis. Some suggested that the labeling of small sections as "problem solving labs" rather than "recitations" might signal progress in this direction.

Instructional Interconnections

Current reform efforts in school mathematics have been influenced by several reports from the National Council of Teachers of Mathematics. The most recent report, *Principles and Standards for School Mathematics* (NCTM, 2000) presents ten standards for mathematics programs in Grades Pre-K to 12. Five standards define the core content (Number, Algebra, Geometry, Measurement and Data Analysis and Probability). Five others define fundamental characteristics of each content area (Problem Solving, Reasoning and Proof, Communication, Representation, and Connections). This report and its predecessor, the *1989 Curriculum and Evaluation Standards for School Mathematics* have generated expectations among many students and teachers that mathematics requires thinking; that mathematics has meaning and application; and that there are multiple approaches to solving problems. Furthermore, since the early 1990s many standards-based school mathematics curricula have been developed and are in use in schools across the United States.

As a result of the increasing spread of the NCTM Standards there is a great need for teachers with deep understanding of this broad range of content. Without such understanding we cannot expect teachers to effectively work with standards-based materials or other challenging mathematics curricula. Several speakers pointed out how at some colleges and universities students seem to be lacking experiences with statistics or probability.

Faculty involved in the preparation of teachers of mathematics should engage in study and discussion of how people learn mathematics.

Much research has been done on how people learn mathematics. Both faculty and graduate students from the disciplines of mathematics and mathematics education—groups that include the instructors of all future K–12 teachers—would better serve prospective mathematics teachers if they were well informed about this research and considered its implications for their own teaching.

Two books were recommended by workshop participants for study: *Knowing and Teaching Elementary Mathematics* (Ma, 1999) and *The Teaching Gap* (Stigler & Hiebert, 1999). Ma describes the "profound understanding of fundamental mathematics" of a group of Chinese elementary teachers, and notes how seldom such understanding was seen in a group of elementary teachers in the United States. Stigler and Hiebert compare and contrast teaching mathematics in Grade 8 in classrooms in Germany, Japan, and the United States using data from the TIMSS video study.

At the time of the workshop, the Mathematics Learning Study Committee of the National Research Council had undertaken a review of research on learning and teaching mathematics in Grades K–8 (NRC, 2001). That review has been published recently, and we, the editors of this report, also recommend it for study.

The Mathematics Learning Study committee demonstrated that a diverse set of mathematicians, psychologists, mathematics educators, teachers and others can work together productively on an issue of national importance. Such collaboration is very important, but not always easy to carry out. Recognizing this leads to our last principle.

Greater communication and cooperation is necessary among all stakeholders in the mathematics preparation of teachers.

Stakeholders in the preparation of mathematics teachers include faculty in mathematics, mathematics education, and education, instructors at two-year and four-year colleges, and teachers and administrators in public and private schools.

One issue that surfaced during the workshop was the language gap between the stakeholders. For example, terms such as deep understanding, algorithm, and proof evoked questions about their meaning and their use. Sometimes the language gap occurred between mathematicians and mathematics educators, sometimes between university and school teachers. Beliefs about what is important sometimes clashed.

Due in part to this gap, the structure and content of preparation programs for mathematics teachers should arise from a partnership between school and university faculty from all departments. Future teachers deserve well-designed programs and courses within departments of mathematics. Efforts to provide such programs to students could be more successful if common objectives for the education of teachers were established, and methods of meeting those objectives were designed, evaluated and revised with participation from all stakeholders.

Mercedes McGowen reported that approximately 40% of teachers in the United States complete some or all of their mathematics and science courses in community colleges. In view of this statistic, two-year college faculty should be invited to participate in and help organize any future national or local discussions about the mathematical preparation of teachers.

More generally, joint seminars and team teaching be used to foster communication across departments involved in mathematics teacher preparation. Such joint seminars might involve concentrated study of some topic in school mathematics. Mathematicians might help mathematics educators gain new insights about the mathematics. Mathematics educators might help mathematicians gain new insights about student learning. Jack Plotkin from the Department of Mathematics at Michigan State University and Dan Chazan from the Department of Teacher Education at Michigan State recently team-taught a mathematics capstone course for majors focusing on the Fundamental Theorem of Algebra. They recommend that team-teaching should also be considered in introductory courses. Communication within mathematics departments can also be fostered by discussions among instructors who teach the same course. These discussions could provide a forum for sharing information about what they do in their sections, and offer opportunities for instructors to consider how one mathematics course builds on what has been learned in other courses.

Economopoulos and Fox described how the North Metro Mathematics Collaborative fosters communication among schools and colleges in suburban Atlanta. This project, centered in the Mathematics Department at Kennesaw State, is dedicated to supporting mathematics teachers in their pursuit of excellence in instruction. Each year it sponsors many professional development courses and workshops in mathematics led by leaders with national stature who attract hundreds of teachers from kindergarten through college. Other means of fostering communication among instructors of mathematics courses for future teachers include: national and regional workshops similar to this one, as well as sessions at national, regional and state meetings, electronic discussion forums, papers in journals, and on-line professional development opportunities.

Participants were particularly interested in discussing the meaning of deep understanding of big ideas such as equivalence, proof, function, and proportional reasoning. Other topics suggested for study were the development and evolution of these ideas in the school mathematics curriculum and how university courses support the development of deep understanding.

Participants frequently mentioned concerns about the lack of resources and support for work in mathematics departments on courses or programs for future teachers. To meet the goals of the MET report on the preparation of future teachers and to implement recommendations such as those reported here, teacher education will need to become a higher priority within mathematics departments.

REFERENCES

Conference Board of the Mathematical Sciences. (2001). *The Mathematical Education of Teachers.* Providence, RI and Washington, DC: American Mathematical Society and Mathematical Association of America.

Ma, L. (1999). *Knowing and teaching elementary mathematics.* Mahwah, NJ: Erlbaum.

Mathematical Association of America. (1991). *A Call for Change.* Washington, DC: MAA.

National Council of Teachers of Mathematics. (1989). *Curriculum and Evaluation Standards for School Mathematics.* Reston, VA: The Council.

National Council of Teachers of Mathematics. (1992). *Professional Standards for Teaching Mathematics.* Reston, VA: The Council.

National Council of Teachers of Mathematics. (2000). *Principles and Standards for School Mathematics.* Reston, VA: The Council.

National Research Council (2001). *Adding it up: Helping Children Learn Mathematics.* J. Kilpatrick, J. Swafford & B. Findell (Eds.). Mathematics Learning Study Committee, Center for Education, Division of Behavioral and Social Sciences and Education. Washington, DC: National Academy Press.

Steen, L., A., Ed. (1990). *On the Shoulders of Giants: New Approaches to Numeracy.* Washington, DC: National Academy Press.

Stigler, J. W. & Hiebert, J. (1999). *The Teaching Gap: Best Ideas from the World's Teachers for Improving Education in the Classroom.* New York: The Free Press.

Usiskin, Z., Stanley, R., Peressin, A. & Marchisotto, E. (2003*). Mathematics for High School Teachers: An Advanced Perspective.* Needham, MA: Prentice Hall.

WORKSHOP PARTICIPANTS

Deborah Ball, Department of Educational Studies, University of Michigan

William Barker, Department of Mathematics, Bowdoin College

Hyman Bass, Departments of Educational Studies and Mathematics, University of Michigan

Susan Beal, Department of Mathematics, St. Xavier University

Charlene Beckmann, Department of Mathematics, Grand Valley State University

Sybilla Beckmann, Department of Mathematics, University of Georgia

Curtis D. Bennett, Department of Mathematics, Bowling Green State University

Dawn Berk, Division of Science and Mathematics Education, Michigan State University

Keely Burns, Department of Mathematics, New Trier High School, Winnetka, Illinois

Joanne Caniglia, Office of Collaborative Education, Eastern Michigan University

Suzanne Chapin, School of Education, Boston University

Michaele Chappell, Department of Secondary Education, University of South Florida

Daniel Chazan, Department of Teacher Education, Michigan State University

Herb Clemens, Department of Mathematics, University of Utah

Christopher Danielson, Department of Mathematics, Michigan State University

Marjorie Economopoulus, Secondary and Middle Grades Education, Kennesaw State University

Joan Ferrini-Mundy, Associate Dean, DSME and Departments of Mathematics and Teacher Education, Michigan State University

Naomi Fisher, MER Forum, Department of Mathematics, University of Illinois - Chicago

Judy Flowers, Department of Mathematics and Statistics, University of Michigan - Dearborn

Marian Fox, Department of Mathematics, Kennesaw State University

Richard Hill, Department of Mathematics, Michigan State University

Stephen Hwang, Department of Mathematics, Michigan State University

Nan Jackson, Department of Mathematics, Lansing Community College

Gary R. Jensen, Department of Mathematics, Washington University

Liz Jones, Department of Teacher Education, Michigan State University

Brian Keller, Department of Mathematics, Michigan State University

Jeannette Kiernan, Navigator Elementary School, Pinckney, Michigan

Karen King, Department of Mathematics, Michigan State University

Michael Lehman, Department of Mathematics, Holt High School, Holt, Michigan

Eric Knuth, Department of Curriculum & Instruction, University of Wisconsin

Valerie Mills, Office of Instructional Services, Ann Arbor Public Schools

Diana Lambdin, School of Education, Indiana University

Perry Lanier, Department of Teacher Education, Michigan State University

Glenda Lappan, Department of Mathematics, Michigan State University

Gary Lewis, Department of Counseling, Educational Psychology, and Special Education, Michigan State University

Mercedes McGowen, Department of Mathematics, William Rainey Harper Community College

Lynn Narasimhan, Associate Dean, Liberal Arts and Sciences, Department of Mathematical Sciences, De Paul University

Ira Papick, Department of Mathematics, University of Missouri

Tom Parker, Department of Mathematics, Michigan State University

Anthony Peressini, Department of Mathematics, University of Illinois

Elizabeth Phillips, Department of Mathematics, Michigan State University

Jacob Plotkin, Department of Mathematics, Michigan State University

Maria Alejandra Sorto, Department of Mathematics, Michigan State University

Sheila Sconiers, COMAP

Sharon Senk, Department of Mathematics, Michigan State University

Kathleen Snook, Department of Mathematics, U. S. Military Academy

Patricia Tinto, School of Education, Syracuse University

Deborah Zopf, Department of Mathematics, Henry Ford Community College

ACKNOWLEDGEMENTS

We wish to thank the following workshop participants for their helpful feedback on earlier drafts of this report: Curt Bennett, Herb Clemens, Gary Jensen, Marjorie Economopolous, and Tom Parker. Thanks also to Bill Berlinghoff, Kevin Clancey, Karen Graham, Libby Krussel, Judy Roitman, Dara Sandow, Annie Selden, and Phil Wagreich for comments made during the Focus Group discussion in New Orleans in January, 2001, and to Richard Askey and Manuel Berriozabal for their thoughtful written comments.

Finally, we would like to thank Lou Anna K. Simon, Provost of Michigan State University, for providing financial support for this workshop, Rebecca Murthum of the Division of Science and Mathematics Education at Michigan State University for making local arrangements, and William Barker, Chair of the CRAFTY subcommittee of the MAA, for the opportunity to host this workshop.

Technical Mathematics: Biotechnology and Environmental Technology[1]

CRAFTY Curriculum Foundations Project
Los Angeles Pierce College, October 5–8, 2000
J. Sargeant Reynolds Community College, October 12–15, 2000

Elaine Johnson, John C. Peterson, and KathyYoshiwara, Report Editors
Bruce Yoshiwara and Gwen Turbeville, Workshop Organizers

Summary

Our fields do not require a lot of advanced mathematics. Basic algebra and statistics are the mathematical topics required for students to complete a biotechnology or environmental Associate of Science (AS) or Associate of Applied Science (AAS) program; environmental AS or AAS programs include some basic geometry and trigonometry as well. Courses in algebra and statistics should use an applications-based approach instead of the traditional textbook approach, considering real-life problems that demonstrate how mathematics is used in the field.

Mathematics should not be a filter that causes students to drop out of our programs. We have found that students do not take remedial courses and then return to the programs. Keep in mind the target population: These students may need more hands-on learning. Students tend to be turned off by dry material. It is important that students learn to use mathematics as a tool, whether or not they "like" mathematics.

To enter the job market, students should master application of the topics listed in Appendices C or D. There is a critical need for more mastery of basics; more depth, less breadth; a need to build systematically on the basics. Mathematics courses should provide the fundamentals. Applications can then be integrated into specific Biotechnology and Environmental Technology courses.

Although there have been major and significant advances in biotechnology and environmental technology, these advances do not dictate the need for changes in the mathematics taught. Advances in technology have not eliminated the need for theoretical understanding. Students should not become too dependent on technology. They should question their answers, using common sense and estimation skills.

However, we do see a need for some changes in mathematics education for our students. Biology is more quantitative than it used to be and experimentation is often based on databases. Statistics must be integrated into the curriculum for technical AAS programs. Students should learn to use databases, manipulate database information, plot and interpret graphs, and present data to others using presentation software packages. More applications and more technology should be incorporated into mathematics courses and there should be more emphasis on teamwork and communication skills.

---

[1] This material is based on work supported by the National Science Foundation under DUE grant no. 0003065. Any opinions, findings, and conclusion or recommendations expressed in this paper are those of the authors and do not necessarily reflect the views of the National Science Foundation.

Narrative

Introduction and Background

This report was compiled by representatives from associate degree programs in biotechnology and environmental technology and from the biomedical and environmental areas. In some cases biomedical engineering and biotechnology have been discussed separately, because, while there is some overlap between the two fields, there are also many differences.

Our fields do not require a lot of advanced mathematics (such as calculus and trigonometry). We agreed that everyone should know more mathematics than is required in their everyday job, but probably not calculus or advanced trigonometry.

Basic algebra and statistics are the mathematical topics required for students to complete a biotechnology or environmental AS or AAS program; environmental AS or AAS programs include some basic geometry and trigonometry as well. Courses in algebra and statistics should use an applications-based approach instead of the traditional textbook approach, considering real-life problems that demonstrate how mathematics is used in the field. Specific topics are listed in Appendices C and D.

Many biotechnology programs do not have a college mathematics requirement. However, chemistry is required for most biotechnology programs and it is common for chemistry to have mathematics requirements. Therefore, the prerequisite mathematics may be driven by chemistry.

There is a strong mandate requiring advanced mathematics, such as calculus, that comes from ABET[2] and articulation requirements with four-year colleges. This calculus requirement also has economic ramifications for institutions. However, there is little need for mathematics past algebra when people enter the workplace. Some trigonometry is needed, but probably not advanced trigonometry. In fact, while biomedical programs require far more mathematics than biotechnology programs, technicians in this field may in fact use less mathematics than biotechnologists.

Most environmental and biotechnology programs integrate mathematics into all their applied laboratory classes. Therefore, there is often applied mathematics instruction in most biotechnology and environmental courses.

General and Philosophical Comments

A question that came up repeatedly in our discussion is who should provide the mathematics instruction for our specific programs? Should mathematics be taught in the technical program itself, or should it be taught by the mathematics department? Should it be taught by "captive" mathematics instructors who work for the technical program?

For economic reasons, and to ensure that courses articulate with four-year schools, students from several technical programs are often combined into a single technical mathematics class. This makes it more difficult for the mathematics teacher to focus on discipline-specific applications. Should students from different programs be combined? We need to look at outcomes to see whether students in combined classes are successfully mastering the material they need.

A related question is how and where to incorporate mathematics into a program. Mathematics should not be a filter that causes students to drop out of our programs. If a student wants to enter a biotechnology program but has poor mathematics skills, and if we send that student to take basic mathematics before beginning program courses, most often we never see the student again.

Students do not take remedial courses and return to the programs. So, what should be the pathway for these students? Can they still move forward in the program? Sometimes students enter programs with poor

[2]ABET is the Accreditation Board for Engineering and Technology, Inc.

backgrounds and become so motivated by the material that they are successful in mathematics. It is important to provide opportunities and not allow mathematics entry requirements to be a barrier.

Many students entering the biomedical engineering technology field have good mathematics skills and/or like mathematics. This ability is what motivates them to enter an engineering technology program. On the other hand, some biotechnology students do not like, and have poor skills in, mathematics. In environmental programs, we have found no consistent attitude toward mathematics. In any case, it is critical that students learn to use mathematics as a tool, whether or not they "like" mathematics.

Understanding and Content

What mathematics is needed in order to complete an applied associate or associate degree and to enter the job market?

A. Biomedical Engineering Technology

Our workgroup unanimously agreed that courses in college algebra and elementary statistics are vital to successful completion of an AAS program. These courses establish the foundation on which technical courses build. The workshop participants in Biotechnology and Environmental Technology have identified specific mathematical topics they considered essential to their areas. These topics are listed by priority in Appendices C and D. (Note: Low priority does *not* mean that understanding of the topic is not needed.)

To enter the job market, students should master application of the topics listed in Appendix C or D. Mathematics courses should provide the fundamentals, and applications can then be integrated into specific Biotechnology and Environmental Technology courses. However, both academia and industry are finding the need to re-teach fundamentals of algebra and statistics in more advanced courses, as well as on the job.

The industry representative listed the following topics necessary for the Associate's Degree, entry into the job market, and career advancement. The referenced problems are in Appendix F.

- Fractions (See problems 1–14)
- Algebra and trigonometric identities (See problem 3)
- Fundamental operations and conversions (See problems 1–14)
- Formula manipulation (See problems 1–14)
- Word problems (See problems 1, 2, 3, 4 and 6)
- Rounding
- Functions and graphs (See problems 1, 2 and 4)
- Linear equations (See problem 5)
- Determinants (See problem 5)
- Trigonometric functions (See problem 3)
    - Angles
    - Right triangles
    - Oblique triangles
- Vectors (See problem 3)
- Quadratic equations (See problem 5)
- Radians and degrees
- Boolean algebra (See problem 6)

Students interested in completing a Bachelor's Degree should also complete coursework in calculus. (Any TAC-ABET accredited Associate's program in Biomedical Engineering Technology must include calculus in its degree requirements.)

B. Biotechnology

Biotechnology technicians must have basic knowledge of many mathematical topics and extensive experience in application. Problems illustrating the topics below can be found in Appendix F.

- Algebra
  - Percentages (See problem 1)
  - Ratios and proportions (See problem 2)
  - Dilutions and concentration calculations (See problems 3, 4 and 5)
  - Measurements
  - Metric system and metric conversions (See problem 6)
  - Significant figures
  - Scientific notation
- Data manipulation and presentation
- Graphing and interpretations of graphs of:
  - Linear equations
  - Quadratic equations
  - Linear regression
  - Exponential/logarithmic functions illustrated in log and semi-log plots (See problems 7 and 8)
  - Construction and interpretation of standard graphs such as those for exponentials and logs (See problem 9)
- Computer software applications
  - Generating and understanding data spreadsheets
  - Graphing software
  - Data base entry and manipulation
- Statistics
  - Sampling, sample size, mean and mode
  - Histograms (See problem 10)
  - Standard deviations (See problem 10)
  - Normal and bimodal distribution (See problem 11)
  - Statistical significance
  - Chi-squares
  - P values
  - Statistical controls (See problem 12)
  - Evaluation of data and measurement systems (See problems 13 and 14)

What mathematics is needed to climb the career ladder?

A. Biomedical Engineering Technology

In biomedical engineering, all workers take advanced service courses. Industry requires engineers to have a baccalaureate or higher degree, but not all technicians need education beyond the AS or AAS level. (This brings up the issue of the transferability of an applied or technical mathematics course to a 4-year institution. This issue is discussed further in response to the next question.)

Graduates tend to move into management in hospitals with the associate degree. They may take business and management courses to advance. Nursing combined with a biomedical degree is very advantageous. Companies really like these people. A graduate who wants to become an engineer will need a BS degree. Thus, moving up the career ladder and getting further education are not necessarily the same.

B. Biotechnology

In biotechnology, employees are often sent back to school to learn what they need to know, such as computers or mathematics. In biomanufacturing, people may be able to advance without further degrees. Bioprocessing technicians will probably need more education in computers; they are less likely to need more mathematics. For R&D (research and development) and for QC/QA (quality control/quality assurance), there is a strong feeling that people need at least a BS degree in order to advance. Therefore, there is pressure for students to take calculus, since it is typically required for a biology BS degree. No one at the workshop could think of a reason why calculus is actually needed, except that it is required for a BS degree. In the work place, statistics and experimental design would be helpful.

In the environmental field, technicians can be employed in jobs ranging from environmental sampler to hazard waste coordinator or environmental manager. However, more education means more upward mobility. There are several 4-year degree programs that have environmental emphasis, including environmental science, environmental biology, environmental chemistry, environmental geology, environmental toxicology, environmental engineering, and environmental management.

Historically, 4-year institutions require a minimum of college algebra and statistics. If mathematics courses were developed for specific applications, they would probably not be transferable towards a baccalaureate degree program. Therefore, a student who is required to complete a non-traditional applied or technical mathematics course for the AS or AAS program will most likely be required to take traditional college algebra and statistics to fulfill the mathematics requirements for a baccalaureate degree program.

What priorities exist among these topics?

Mastery of certain skills is critical. This includes performing calculations related to percentages, dilutions, solutions, concentration, exponents, scientific notation, and basic algebra. Students must practice these skills over and over again. Mastery of the basics is more important than exposure to a lot of mathematics.

Students must be able to extract information from written materials, both obtaining needed data and deducing appropriate methods for solving the posed problems. They must be able to apply what they have read to their own problems. They must be able to solve word problems. They must be able to extract numerical information in the laboratory and solve problems using that information. As people advance up a career ladder they may need more mathematics skills, particularly if they move into research or engineering.

Specific topics relevant to Biotechnology and Environmental Technology are prioritized in Appendix C. The workshop participants identified these topics and ranked them according to the importance of their applications.

Examples:

1. Situation: A laboratory technician collects water samples and must test the samples for their chloride content. The technician needs to understand the concepts of random versus non-random sampling and the relevance of sample size on the interpretation of the testing results.
2. The testing requires a chloride standard to be diluted several ways in order to test various known concentrations of the standard using a defined test method. The technician needs to understand fractions and accuracy of measurements.
3. The standards and the samples are analyzed using the defined test method and a standard curve is obtained for the various known concentrations of the chloride standard. The technician needs to know how to plot data, recognize the shape of the curve, identify outliers and apply linear regression analysis.
4. The chloride content of the samples (unknown) is determined from the standard curve. The technician needs to know the relationship between x and y (how to use the linear regression formula).

What is the desired balance between theoretical understanding and computational skill? How is this balance achieved?

The student's first priority is to solve applied problems in the laboratory. Minimal theory is required, such as is taught in high school algebra. If a person moves into research, then he or she will probably need more theory. In production, technicians need only very simple multiplication and division skills.

Technology

The technician must be able to solve numerical problems using databases, use the Internet to connect to national databases, and manipulate database information. For example, the technician might look at correlations or percent differences in genome sequences.

In general students should be able to use ordinary and statistical calculators and should become comfortable with spreadsheets for calculation and graphing. However, they should not become too dependent on technology. They should question their answers, not accept them blindly. Students should use common sense and estimation skills.

Students should understand software associated with equipment; programming instruments is important. There may be statistics associated with using instruments.

Future Trends

With proteomics and genomics, students need more instruction in information sciences. There has been rapid growth in bioprocessing. In Genentech, for example, there are ten jobs in bioprocessing for each job in R&D. Technicians need to be comfortable with some engineering. They may need some electronics and knowledge of HVAC systems. They need to know how to fix and operate instruments. The ability to operate instruments is more important for AAS graduates than knowing the mathematics and engineering behind the instruments. Twenty years ago technology became hugely important in nursing, and people needed to learn about this technology, so they invented a field to create liaisons between technology and clinicians. Robotics is now becoming a big issue. Technicians need to be able to install instruments without instruction.

Instructional Techniques

What are the effects of different instructional methods in mathematics on students in your discipline?

Much of this has been discussed above. Keep in mind the target population: These students may need more hands-on learning. Students tend to be turned-off by dry material.

What instructional methods best develop the mathematical comprehension needed for your discipline?

In addition to what was discussed above, real life examples, verbal examples of logic (valid and invalid reasoning), and hands-on use of technology are most effective.

Instructional Interconnections

What changes have occurred in your discipline that should affect what mathematics is taught?

Although there have been major and significant advances in biotechnology and environmental technology, these advances do not dictate the need for changes in the mathematics taught. Advances in technology have not eliminated the need for theoretical understanding. Unfortunately, advances in technology have reduced the requirement for computational skills, and it is apparent that the youth of today lack even the most basic computational skills (multiplication and division), due to the widespread use of calculators in primary and secondary schools. The widespread use of computerized equipment in society further adds to the problem. This is doubly unfortunate in that the application of manual computational skills can, in certain situations, reinforce theoretical understanding.

We recommend that mathematics faculty coordinate with technology faculty on a periodic basis to keep abreast of any changes that might affect what mathematics topics are taught. In particular, we see the following eight changes that should affect what mathematics is taught.

1. Volumes are usually small with units such as micro (microliter), nano (nanoliter), pico (picoliter), but can include 1000s or kilos (kiloliters). Small and large scales affect mathematics.
2. Bioinformatics and computer technologies have rapidly changing needs in mathematics skills.
3. Biology is more quantitative than it used to be.
4. Experimentation is often based on databases.
5. People tend to move more across different disciplines.
6. Regulatory affairs have increaded the need for more familiarity with statistics.
7. Teamwork is more important in the workplace.
8. There may be more acceptance in the profession of people with skills instead of degrees.

What changes have occurred in your discipline that should affect how mathematics is taught?

Advances in technology, computer programs, and equipment in our discipline drive the need for using software programs and databases such as *Lotus 123*, *Oracle*, *Excel*, etc. in the classroom. These are tools commonly used in industry so our students need to be exposed to them during their education. We see the following seven changes that should affect how mathematics is taught.

1. More teamwork in solving mathematical problems, since there is a need in the workplace for this skill.
2. More oral communication/presentation skills required to present work, such as graphs.
3. More use of computers to present information, to get information (including use of Internet) and to plot data.
4. Interpretation of graphs, ability to plot or show data in many different ways.
5. Ability to make computer presentations means students are expected to display data in many ways, using computers to make such presentations understandable.
6. System-level troubleshooting is now more important than individual instruments. It often costs too much to fix an individual instrument, so technicians just replace the whole system. Boards are cheaper and complex, so technicians don't trouble-shoot them.
7. Technicians need enough computer literacy skills to learn new software packages, so students should use a variety of software packages in mathematics classes.

The group disagreed on whether it is more important to be flexible in using various software packages, or to know specific commonly used software packages. It might be helpful for mathematics instructors to use whatever packages are most common in biotechnology programs.

What changes are needed in the mathematics curriculum in order to satisfy the needs of AAS students and technicians?

1. Statistics must be integrated into the curriculum for technical AAS programs.
2. The biggest need is for more applications and more technology incorporated into mathematics courses.
3. Word application problems need to be used more to supplement the topics taught.
4. There is a critical need for more mastery of basics; more depth, less breadth; a need to build systematically on the basics. Mastery requires repetition, and is different than exposing student to lots of material. Mastery also involves making sure students can apply mathematics in their own field.

5. Students should learn how to present data to others using presentation tools (computer tools) and speaking skills.
6. Estimation in various situations should be taught. For example, estimate how many basketballs would fit in a room? How long will it take to get enough material to sequence a gene? Estimation helps students know if their results make any sense when they use a calculator or computer.
7. Many students go into technical fields because they are hands-on people. Teachers need to respond to students who may be very kinesthetic and to make instruction hands-on and interesting.
8. Teachers should use projects involving multiple mathematics skills that also tie in technical content. For example, figure out a correlation between sequence and function based on genome database information, or calculate enzyme activity based on laboratory data.
9. Changes in the assessment of student competencies should also be considered.

What instructional methods might mathematics instructors use to develop or reinforce non-mathematical skills or understandings in your discipline or company?

1. Use a lot of teamwork in classes.
2. Have students bring in their own problems with their own data, particularly if working with older students who have more background. Have students find data from journals or publications. Have students find the ways in which mathematics is applied.
3. Reinforce writing skills and deductive reasoning by requiring students to write an explanation of how they arrived at their result or answer and how they interpreted their results.
4. Ask students to predict a non-exact result (example: given a problem, will the result be greater than or less than a value).
5. Teach students while they are sitting in a circle; this arrangement sets up a different dynamic.
6. Let students practice teaching themselves; use peer mentoring.

How can dialogue on educational issues between your discipline and mathematics best be maintained?

Maintain an open line of communication through planned classroom visitations (guest speakers from industry, AS or AAS program faculty), conduct surveys and hold local workshops with industry representatives and AS or AAS program faculty. Here are nine ways in which this can be accomplished:

1. Have industry representatives help mathematics teachers become aware of the technical jobs available in their communities and the mathematical needs of these positions.
2. Have advisory groups from industry or technical faculty work with mathematics teachers, have mathematics teachers on biotechnology program faculty.
3. Acquaint mathematics teachers with work keys, SCANS, and occupational skill standards.
4. Have discipline-specific people present talks at mathematics meetings.
5. Have mathematicians and biotechnologists work on more complex project-based problems. Problems could involve genomic information, or interesting applications from news.
6. Bio-Link and other biotechnology instructors could act as a bridge to help mathematics teachers see real world applications, for example, in the human genome project. There could be collaboration between AMATYC and Bio-Link.
7. Encourage team teaching between mathematics and technical program teachers.
8. Have guest lecturers come to talk about how mathematics is used in their profession.
9. Have a mathematics teacher go to industry and review the mathematics currently being used in companies.

REFERENCES

Cincinnati State Technical and Community College Catalog, 3520 Central Parkway, Cincinnati, Ohio 45223.

Seidman, Lisa A. and Cynthia J. Moore. Basic Laboratory methods for Biotechnology: Textbook and Laboratory Reference. Upper Saddle River, NJ: Prentice Hall, 2000.

U.S. Department of Labor, Bureau of Labor Statistics. Occupational Outlook Handbook Online. `www.bls.gov`

WORKSHOP PARTICIPANTS

Brenda Breeding, Professor of Biology, Oklahoma City Community College, Oklahoma City, Oklahoma.

Daniel R. Brown, Ph.D., Lead Instructor of Microbiology and Biotechnology, Santa Fe Community College, Gainesville, Florida.

Sybil Chandler, Lead faculty for the Environmental Health and Safety Technology Associate Degree program of Metropolitan Community College's (MCC) Maple Woods Community College in Kansas City, Missouri.

Lois Dinterman, Director of Bioprocessing for Biolex, Inc., Pittsboro, North Carolina.

Linnea Fletcher, Ph.D., Austin Community College, Austin, Texas, and South Central Regional Director for Bio-Link

Leland Grooms, adjunct faculty and advisory committee member who assisted in the development for Environmental Sciences and EHS for Haskell Indian University, Lawrence, Kansas and Maple Woods Community College, respectively.

Elaine Johnson, Ph.D., City College of San Francisco, San Francisco, California, and National Director for Bio-Link, an Advanced Technological Education (ATE) Center of Excellence of the National Science Foundation

Wendie Johnston, Ph.D., Pasadena City College, Pasadena, California: Program Director and EdNet Director

JoAnne Marzowski, Ph.D., Bristol Myers Squibb, Oncogene, Seattle, Washington: Researcher in molecular biology

Donald McAfee, Ph.D., Chair and CEO of Discovery Therapeutics, Inc., Richmond, Virginia, a pharmaceutical drug discovery and drug developing therapeutics for cardio-renal indications.

Jim McGillem, Biomedical, Facility Director, St. Mary's Warrick health care facility, Boonville, Indiana

Lisa Seidman, Ph.D., Madison Area Technical College, Madison, Wisconsin: Biotechnology Program and North Central Regional Director for Bio-Link

William H. Woodruff, Department Head, Biotechnology Department, Alamance Community College, Graham, North Carolina, and Southeast Regional Director for the BioLink Advance Technology Bio-Link Advanced Technological

Steve Yelton, Cincinnati State Technical and Community College, Cincinnati, Ohio: Program Chair, Electronics, Biomedical Electronics & Computer Network Engineering Technology,

ACKNOWLEDGEMENTS

The members of the panels would like to thank Bruce Yoshiwara, of Los Angeles Pierce Community College and Gwen Turbeville of J. Sargeant Reynolds Community College for coordinating the workshops at their respective campuses. We would also like to thank the Mathematical Association of America for initiating this series of workshops and the American Mathematical Association of Two-Year Colleges for providing this opportunity to consider and make recommendations on a topic of importance to the information technology community, especially project directors Mary Ann Hovis, Rob L. Kimball, and John C. Peterson. Finally, we would like to thank the National Science Foundation for providing the funds to support these meetings.

APPENDIX A: Definitions

The following definitions will help those who are not familiar with the fields of biotechnology and environmental technology to understand our report more fully.

Bioinformatics: A field of science formed by the merging of biology, computer science, and information technology.

Bio-Link: An Advanced Technological Education (ATE) Center of Excellence of the National Science Foundation.

Bioreactor: A piece of equipment used to produce products such as pharmaceuticals using live organisms.

Prokaryotic organisms: Single-cell organisms, such as bacteria, that lack a true membrane-bound nucleus and whose DNA is usually a single molecule.

Eukaryotic organisms: Organisms such as plants and animals made up of cells that contain a true membrane-bound nucleus.

Bioprocessing: The manufacturing of products using living cells.

Genome: The total complement of genes in an organism.

DNA: Deoxyribonucleic acid: a polymer that is part of chromosomes that contain the genetic code.

SCANS: The Secretary's Commission on Achieving Necessary Skills for the workplace.

APPENDIX B: Description of Programs

In this document we discuss people preparing to work in one of three major areas: (a) Biotechnology, (2) Biomedical Electronics Engineering Technology, or (3) Environmental Health and Safety.

Biotechnology

Biotechnology is a composite of many technical fields, and graduates of biotechnology AS or AAS programs can choose from many different jobs. For example, one graduate may be working on designing a new bioreactor, another may be developing a new assay, and still another may work in production or quality control. Therefore, the breadth and depth of the mathematical expertise one needs is determined by the specific technical field.

A general description of biotechnology programs follows: Biotechnology programs provide education for entry-level technicians in the biotechnology industry. The steady growth of biotechnology and related industries has resulted in the demand for highly skilled technicians. Biotechnology programs prepare skilled technicians to work at the entry level in a wide variety of scientific fields, including: research and discovery laboratories, service and quality assurance laboratories, food, water, soil and product testing laboratories and manufacturing facilities. Skills necessary to support these activities include knowledge of regulatory affairs, tissue culture of plant and animal cells, production with bacterial, yeast, mammalian and plant cells, molecular biology techniques, quality assurance, business, electronics, technical writing, web management, library science, computer science, chemistry, biochemistry, biology, microbiology, physics, analytical laboratory techniques, laboratory instrumentation, growth, isolation and characterization of prokaryotic and eukaryotic organisms, histologic techniques, and immunological techniques.

Biomedical Electronics Engineering Technology

From the catalogue of Cincinnati State College:

> The Biomedical Electronics Engineering Program (BMET) was created because of the need for technicians who repair, maintain, modify and design complex medical instrumentation. This person is employed in hospitals as well as medical equipment manufacturers. The BMET graduate will have advanced electronic skills as well as education in the following areas:
>
> installation and calibration of biomedical equipment
>
> operation of safety and maintenance programs
>
> The biomedical electronics technician is a professional whose broad background in electronics and instrumentation will make the graduate an asset to any organization.

* * * * *

Starting salaries for BMET graduates are about \$30–35,000 in biomedical engineering. Graduates have an opportunity to make \$70,000 if they can repair sophisticated medical, imaging, or laboratory equipment, but some of these jobs require a lot of traveling. Later, when they are tired of traveling, job opportunities may exist in a hospital although the hours are not ideal. However, \$40–50K job opportunities may exist with better working hours.

Environmental Health and Safety

Environmental Health and Safety provides a foundation in aspects of environmental health and safety technology including emergency response planning, OSHA, EPA, and DOT standards and legislation, air- and water-quality management, accident and incident investigation, characteristics and hazards of hazardous material, and working on hazardous waste sites. In addition, emphasis is placed on managing chemical and biological substances and studying their effects on the environment. The AAS degree program includes additional course work in chemistry, biology and physics.

The median salary of inspectors and compliance officers in environmental health and safety was over $34,000 in 1994, with 10% of these jobs commanding salaries of about $60,000.

Environmental health and safety careers offer good salary and opportunity. According to Dept. of Labor statistics, employment in the environmental, health and safety fields will grow through the year 2005, spurred by public demand for a livable environment, safe working conditions and non-hazardous consumer products.

There is a wide range of employment opportunity. The primary employers are industry, environmental consultants, and all levels of the government. Specialists are in demand in a wide range of fields including agriculture, aviation, electronics, health care, lumber, manufacturing, municipalities, park systems, petroleum, and local and Federal government.

Description of Students

We maintain open entry into almost all our technical programs; therefore students come in with very diverse mathematics backgrounds and abilities. In some programs a large percentage of the students enter with BS degrees, while other programs have few such students. There is still a gender difference. Biotechnology programs may have more women; environmental and engineering may have more men. The average age in community college programs is about 28, and many people are retraining.

APPENDIX C: Biotechnology

References:

Basic Laboratory Methods for Biotechnology
Textbook and Laboratory Reference
ISBN: 0-13-795535-9

Technical Math
Robert Smith
ISBN: 0-8273-6808-9
Algebra application examples

3= high; 2=medium; 1=low

Subjects		Priority		
		1	2	3
Basic Algebra				
1.	Estimating the answer/Approximating/Fuzzy Math			X
2.	Fractions			X
3.	Decimals			X
4.	Percents, Percentiles Rank, Interpretation of Percentile Ranks			X
5.	Graphs, Plotting, Bar and Lines			X
6.	Creating Lines, Graphs, Thresholds			X
7.	Measurements: Precision, Accuracy, and Tolerance			X
8.	Measurement Unit (length, mass, volume, distance, time, temperature)/Unit Factorization (conversation)			X
9.	Metric and English conversation			X
10.	Word application problems			X
11.	Scales of Measure (nominal, ordinal, interval, ratio)			X
12.	Algebraic Expressions			X
13.	Signed numbers			X
14.	Powers and Roots			X
15.	Basic Algebraic Operations			X
16.	Linear and Quadratic Equations			X
17.	Formula and variation problems			X
18.	Ratio and Portions (Variation problems)			X
19.	Cartesian Coordinate System and Graphs of Linear Equations			X
20.	Systems of Equations			X
21.	Dilutions			X
22.	Molarity (Normality), pH			X
23.	Log functions and graph			X
24.	Valid and invalid reasoning			X
25.	Geometry & Trigonometry	X		
26.	Measurement of diameter, area	X		
27.	Calculation of volume	X		
28.	Pythagorean Theorem	X		

	Subjects	Priority 1	2	3
Basic Statistics (Applied Statistics)				
29.	Variables and Constants (continuous and discrete variables)			X
30.	Qualitative, quantitative, symbolizing variable in computations			X
31.	Tabulating Data (frequency distributions)			X
	Graphing Data (discrete & continuous data)			
32.	Histogram			X
33.	Frequency Polygon	X		
34.	Smoothed and Misleading Graphs	X		
35.	Describing the Shapes of Distributions			X
	Measures of Central Tendency			
36.	Mode	X		
37.	Median			X
38.	Mean			X
39.	Properties of the Mean			X
40.	Measures of Variance (range, variance, standard deviation, relative deviation)			X
41.	Data Collection and graphing of collected data			X
42.	Considerations in Selecting a Measure of Central Tendency			
43.	Scale of Measurement			X
44.	Shape of Distribution		X	
45.	Normal Distribution (normal curve in both directions, distribution table, interpretation of scores, normal curve equivalents, T-scores)			X
46.	Sets (union, intersection, compliment), definition of probability, conditions, basic laws, such as A or B, not A, A and B, B given that A is a curve, independent and dependent event, mutual exclusive events Simple Binomial probability (success/failure) Basic understanding: high/ Application: low			X
	Measures of Relationship			
47.	Characteristics of Associations Between Variables (Scatter Diagram, degrees and directions of association)			X
48.	Calculating a Correlation Coefficient (for example: Pearson Product Moment, Interpreting a Correlation Coefficient			X
49.	Correlation/causation			X
50.	Scale of the Pearson *r*			X
51.	Range of *X*- and *Y*-variables			X
52.	Prediction and Simple Linear Regression			X
53.	Simple Linear Regression			X
54.	The Standard of Error of Estimate			
55.	Marginal and Conditional Distributions			X
56.	Samples and Estimation			X

	Subjects	Priority 1	2	3
	Type of Samples			
57.	Non random Sample	X		
58.	Probability Sample		X	
59.	Sampling distribution (theory only)			X
60.	Central Limit Theorem		X	
61.	Confidence intervals			X
62.	Accuracy, bias, and precision in sampling			X
63.	Obtaining a confidence interval with a t-distribution			X
	Testing Hypotheses			
64.	General Steps			X
65.	Possible Errors of Statistical Decision: Type I and Type II		X	
	t-distribution and t-test			
66.	relationship between the z- and t- statistics		X	
67.	degrees of freedom		X	
68.	characteristics of the t-distribution		X	
69.	testing hypotheses about means-using the t-distribution		X	
70.	testing the hypothesis μ = some value		X	
71.	testing difference between means		X	
72.	differences between means for independent random samples from equally variable populations differences between means for dependent samples			X
73.	Chi-Square Distribution, Goodness of Fit, Independence (general case, 2×2 contingency table)			X

APPENDIX D: Environmental Technology

References:

Basic Laboratory Methods for Biotechnology
Textbook and Laboratory Reference
ISBN: 0-13-795535-9

Technical Math
Robert Smith
ISBN: 0-8273-6808-9
Algebra application examples

3 = high; 2 = medium; 1 = low

	Subjects	Priority 1	Priority 2	Priority 3
Basic Algebra				
1.	Estimating the answer/Approximating/Fuzzy Math			X
2.	Fractions			X
3.	Decimals			X
4.	Percents, Percentiles Rank, Interpretation of Percentile Ranks			X
5.	Graphs, Plotting, Bar and Lines			X
6.	Creating Lines, Graphs, Thresholds			X
7.	Measurements: Precision, Accuracy, and Tolerance			X
8.	Measurement Unit (length, mass, volume, distance, time, temperature)/ Unit Factorization (conversation)			X
9.	Metric and English conversation			X
10.	Word application problems			X
11.	Scales of Measure (nominal, ordinal, interval, ratio)			X
12.	Algebraic Expressions		X	
13.	Signed numbers			X
14.	Powers and Roots			X
15.	Basic Algebraic Operations			X
16.	Linear and Quadratic Equations			X
17.	Formula and variation problems			X
18.	Ratio and Portions (Variation problems)			X
19.	Cartesian Coordinate System and Graphs of Linear Equations			X
20.	Systems of Equations			X
21.	Dilutions			X
22.	Molarity (Normality), pH			X
23.	Log functions and graph		X	
24.	Valid and invalid reasoning			X
25.	Geometry & Trigonometry		X	
26.	Measurement of diameter, area		X	
27.	Calculation of volume		X	
28.	Pythagorean Theorem	X		

	Subjects	Priority 1	2	3
Basic Statistics (Applied Statistics)				
29.	Variables and Constants (continuous and discrete variables)			X
30.	Qualitative, quantitative, symbolizing variable in computations			X
31.	Tabulating Data (frequency distributions)			X
	Graphing Data (discrete & continuous data)			
32.	Histogram			X
33.	Frequency Polygon	X		
34.	Smoothed and Misleading Graphs	X		
35.	Describing the Shapes of Distributions			X
Measures of Central Tendency				
36.	Mode			X
37.	Median			X
38.	Mean			X
39.	Properties of the Mean			X
40.	Measures of Variance (range, variance, standard deviation, relative deviation)			X
41.	Data Collection and graphing of collected data			X
	Considerations in Selecting a Measure of Central Tendency			
42.	Scale of Measurement			X
43.	Shape of Distribution			X
44.	Normal Distribution (normal curve in both directions, distribution table, interpretation of scores, normal curve equivalents, T-scores)			X
45.	Sets (union, intersection, compliment), Definition of Probability, conditions, basic laws, such as A or B, not A, A and B, B given that A is a curve, independent and dependent event, mutual exclusive events Simple Binomial probability (success/failure) Basic understanding – high/ Application – Low			X
46.	Measures of Relationship			
47.	Characteristics of Associations Between Variables (Scatter Diagram, degrees and directions of association)			X
48.	Calculating a Correlation Coefficient (for example: Pearson Product Moment, Interpreting a Correlation Coefficient			X
49.	Correlation/causation			X
50.	Scale of the Pearson r			X
51.	Range of X- and Y- variables			X
52.	Prediction and Simple Linear Regression			X
53.	Simple Linear Regression			X
	The Standard of Error of Estimate			
54.	Marginal and Conditional Distributions			X
55.	Samples and Estimation			X

	Subjects	Priority		
		1	2	3
56.	Type of Samples			
57.	Non random Sample	X		
58.	Probability Sample		X	
59.	Sampling distribution (theory only)			X
60.	Central Limit Theorem		X	
61.	Confidence intervals			X
62.	Accuracy, bias, and precision in sampling			X
63.	Obtaining a confidence interval with a t-distribution			X
	Testing Hypotheses			
64.	General Steps			X
65.	Possible Errors of Statistical Decision: Type I and Type II		X	
	t-distribution and t-test			
66.	relationship between the z- and t-statistics		X	
67.	degrees of freedom		X	
68.	characteristics of the t-distribution		X	
69.	testing hypotheses about means-using the t-distribution		X	
70.	testing the hypothesis μ = some value		X	
71.	testing difference between means			X
72.	differences between means for independent random samples from equally variable populations differences between means for dependent samples			X
73.	Chi-Square Distribution, Goodness of Fit, Independence (general case, 2×2 contingency table)		X	

APPENDIX E: Sample Biomedical Problems

1. A patient is connected to a medical device via two electrodes as shown below, each with a diameter of 10 cm. The maximum allowable leakage current is 0.005 amps (5 milliamps). Calculate the maximum voltage allowed across the electrode leads to maintain a safe current for the following conditions: (a) dry skin, (b) gel coated skin, and (c) penetrated skin. Assume $R_V = 200\,\Omega$. Voltage = $I(R$ electrode $1 + R_V + R$ electrode 2).

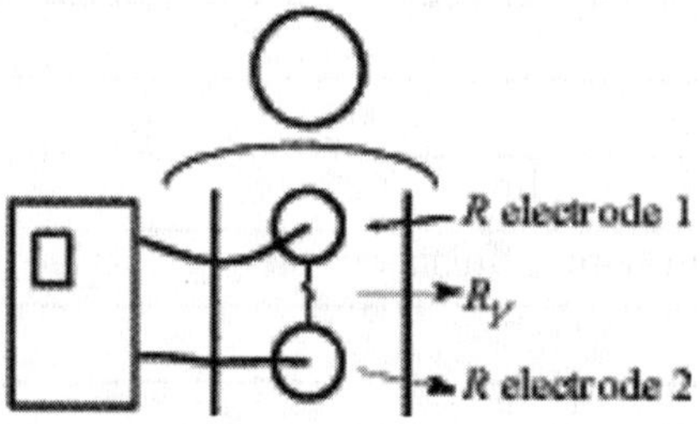

2. A device is brought into the lab that has no output. There are no service manuals available for this device. Upon further investigation, you find a burned out capacitor, the value is unreadable. You find that the circuit looks like the one shown below. What should be the value of C to maintain less than 2 mV of ripple?

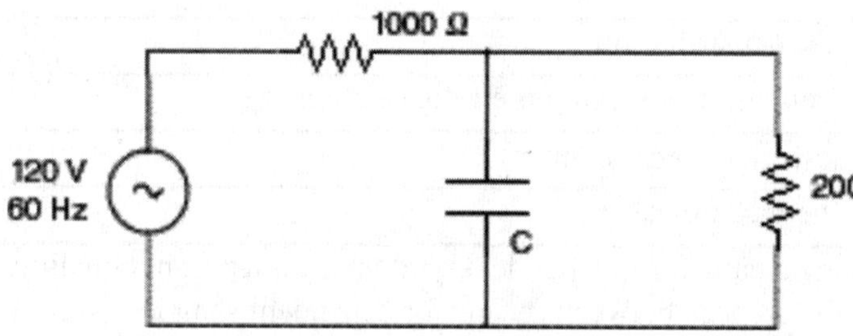

3. A transducer is intended to be interfaced to a measuring device. The transducer produces a minimum sinusoidal output of 0.05 mV @ μA. The measuring device requires a minimum input of 10 mV. Design an amplifier that will perform this interface. Be sure to specify component types including part numbers.

4. You are asked to design a resistive load to test output of a DC power supply. The power supply must be capable of producing 2 amps at 100 volts. The load must be designed with a 200% safety factor. Specify the value of the resistive load including resistance and power ratings.

5. Using the diagram below, calculate currents I_{Total}, I_{R_2}, and I_{R_3}, voltage drop across R_4, and the power dissipated across R_3.

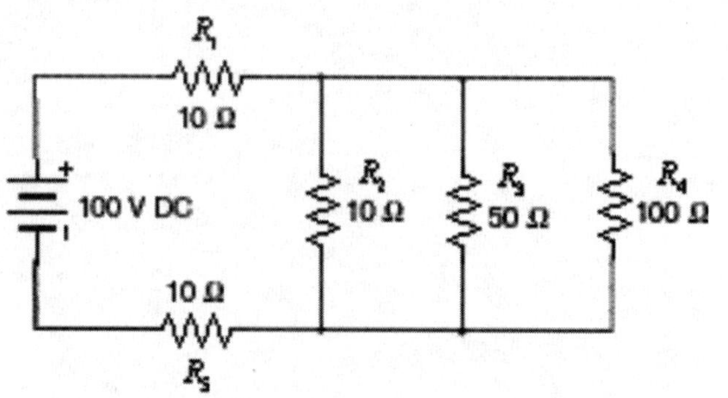

6. An equipment cabinet must be protected by safety interlocks. There are two access panels with interlock switches. There is a key switch and there is a start button. When the access panel is open, the switch outputs a logic level 1, when closed, a 0. The only time that the process will start is when all panels are closed and the key switch is on. The key switch outputs a 1 in the on position. When these conditions are met, output a 1 to the "start process" line. If conditions are not met, output a 1 to the "system halted" line.

   Generate a truth table to describe the process and design a circuit to produce the desired effect. The circuit should be the most simplified possible using Boolean algebra.

APPENDIX F: Sample Biotechnology Problems[3]

1. There are about 3×10^9 DNA base pairs in the human genome. Human chromosome 21 is the smallest chromosome (besides the Y chromosome) and contains about 2% of the human genome. About how many base pairs comprise chromosome 21?

2. If there are about 1×10^2 blood cells in a 1.0×10^{-2} mL sample, then about how many blood cells would be in 1.0 mL of this blood?

3. If the concentration of magnesium sulfate in a solution is 25 g/L, how much magnesium sulfate is present in 100 mL of this solution?

4. Suppose you have 20 μL of an expensive enzyme and you cannot afford to purchase more. The enzyme has a concentration of 1000 units/mL. You are going to do an experiment that requires tubes with a concentration of I unit/mL of enzyme and each tube will have 5 mL total volume. How much enzyme does each tube require? How many tubes can you prepare before you run out of enzyme?

5. A stock solution initially has a concentration of 20 mg of solute per liter. A diluted solution is prepared by removing 1 mL of stock solution and adding 14 mL of water. What is the concentrate of solute in the diluted solution? How much solute is present in 1 mL of the diluted solution?

6. Suppose bacteria are growing in a flask. The growth medium for the bacteria requires 5 g of glucose per liter. A technician has prepared some medium and added 0.24 lb of glucose to 25 L. Did the technician make the broth correctly?

7. Cells in culture are treated in such a way that they are expected to take up a fragment of DNA containing a gene that codes for an enzyme. The activity of the enzyme in cells that take up the gene can be assayed. The more active the enzyme, the better. Suppose a researcher isolates 45 clones of treated cells and measures the enzyme activity in each clone. The results are shown in activity units.
   a. Find the range, median, mean, and standard deviation for the data from these 45 treated clones.
   b. Plot these data on a histogram and show on the plat where the mean and median are located.
   c. Do you think the cells have taken up the gene fragment containing the enzyme, based on these data? Explain.
   (45 data values follow)

8. A technician customarily counts the number of leaves on cloned plants. The results of nine such counts in successful experiments are:

   75 54 55 61 71 67 51 77 71

   If the technician obtains a count of 79, is this a cause for concern?

   Perform statistical calculations to determine whether 79 leaves is out of the range of two standard deviations.

9. Consider a radioactive solution that has a half life of 1 hour and an activity of 400 disintegrations/minute initially. Figure 10.21(a) shows this phenomena graphed on a normal rectangular graph and

[3] All examples in Appendix F are taken from Seidman and Moore's. *Basic Laboratory methods for Biotechnology: Textbook and Laboratory Reference.*

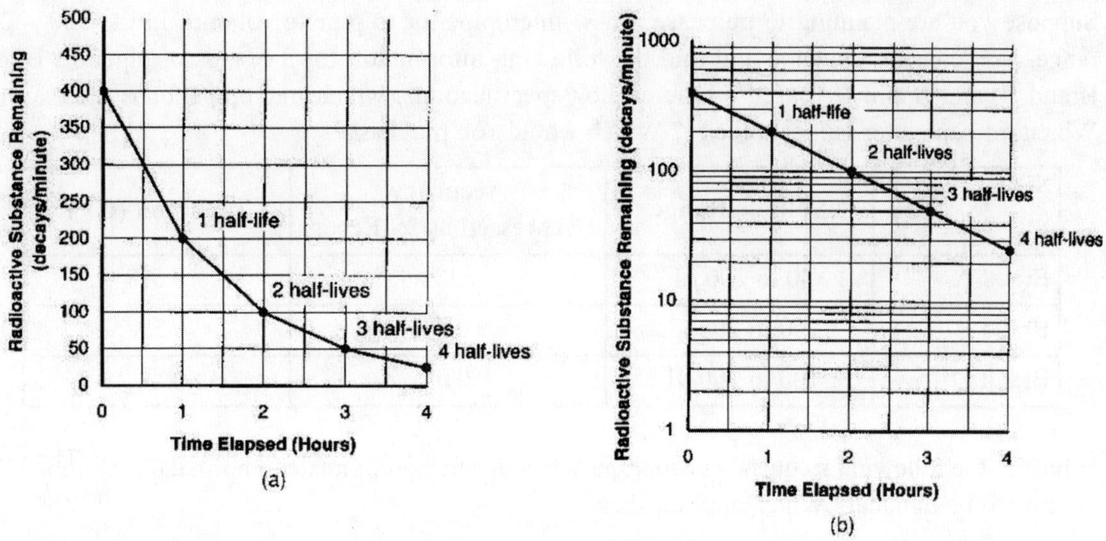

Figure 10.21. The Relationship Between Time Elapsed and Radioactivity Remaining. a. The relationship on normal graph paper. **b.** The relationship plotted on semilog paper.

Figure 10.21(b) shows the same data graphed on semilog paper. Discuss the advantages and disadvantages of each graph.

10. It is difficult to kill all microorganisms in a material. A solution or material to be sterilized is typically heated under pressure. The following graph shows the effect of time of exposure to heat and bacterial death (based on information from *Principles and Methods of Sterilization in the Health Sciences*, John J. Perkins. 2nd edition, Charles C. Thomas, Springfield, 1983).

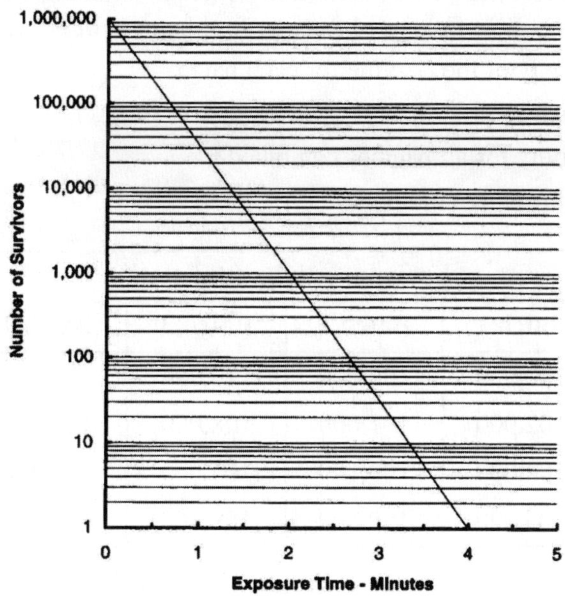

a. About how many bacteria were there at the beginning of the experiment, before exposure to heat?

b. About how many bacteria were present after 2 minutes of treatment?

c. Why did the investigators show their data on a semilog plot?

11. Suppose you are planning to purchase a new micropipettor to pipette volumes in the 100 – 200μL range. You consult a catalog and find the following information for 3 brands of pipettor: Brand A, Brand B, and Brand C. Based on the catalog specifications, which micropipettor is most accurate? Which micropipettor is most precise? Which would you purchase?

	Volume Range	**Accuracy (Expressed as % Error)**	**Precision (CV)**
Brand A	40 to 200 μL	±1%	0.5%
Brand B	100 to 200 μL	±0.5%	0.3%
Brand C	100 to 200 μL	±0.3%	0.4%

12. Which of the following frequency histograms most closely approximates a normal distribution? Which appears to be bimodal? Which appears skewed?

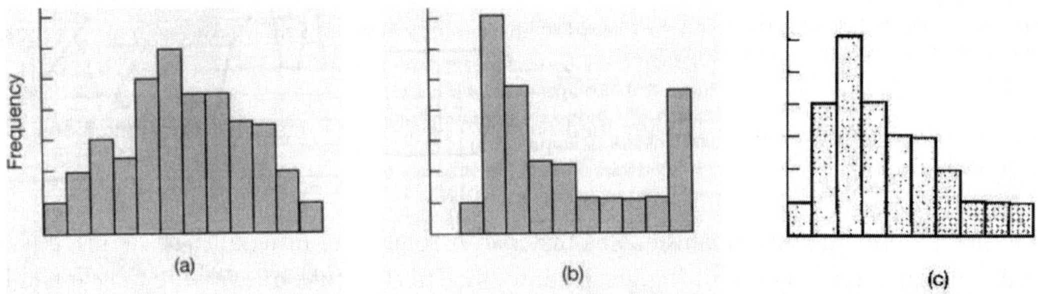

13. A biotechnology company manufactures a particular enzyme that is used to cut DNA strands. The enzyme's activity can be assayed and is reported in terms of "units/mg". Each batch of enzyme is tested before it is sold. The results of repeated tests on four batches of enzyme are shown in the table.

a. What is the mean activity of the enzyme for each batch?

b. What is the SD for each batch?

c. What is the mean activity for all batches combined?

d. What is the SD for all batches combined?

Enzyme Activity (units/mg)			
Batch 1	**Batch 2**	**Batch 3**	**Batch 4**
100,900	100,800	110,000	123,000
102,000	101,000	108,000	121,000
104,000	100,100	107,000	119,000
104,100	100,800	109,100	121,000

Technical Mathematics:
Electronics, Telecommunications, and Semiconductor Technology[1]

CRAFTY Curriculum Foundations Project
Los Angeles Pierce College, October 5–8, 2000
J. Sargeant Reynolds Community College, October 12–15, 2000

Bob L. Bixler, James Hyder, John C. Peterson, and Kathy Yoshiwara, Report Editors
Bruce Yoshiwara and Gwen Turbeville, Workshop Organizers

Summary

All three of these professions work in the electronics area, but the level of mathematics needed by the different fields varies considerably. Semiconductor technicians need little more than ratios, elementary algebra and elementary statistics. Electronics and telecommunications technicians need more mathematics, including complex numbers, trigonometry, elementary geometry, elementary statistics, and differential and integral calculus. All areas want their students to be able to use technology, such as spreadsheets, to analyze mathematics.

Better communication between mathematics and technology departments was considered vital. Suggestions included having mathematics instructors take courses in technical areas to gain a better feel for how mathematics is used, and having technical mathematics faculty be members of technology department advisory boards.

Students need more instruction on how to learn and analyze: scenarios, word problems, and open-ended situations should be used. They need to know which type of equation to select as the method for solving a particular problem. Team projects and collaborative learning activities are essential. Teaching with calculators might be a disservice if students meet only computers in the workplace.

Narrative

Introduction and Background

Technicians are essentially troubleshooters and repair persons. Some work in the field repairing equipment such as copy machines while others work at a test bench. Some work for manufacturing companies repairing assembly equipment while others work in engineering laboratories.

[1] This material is based on work supported by the National Science Foundation under DUE grant no. 0003065. Any opinions, findings, and conclusions or recommendations expressed in this paper are those of the authors and do not necessarily reflect the views of the National Science Foundation.

Sometimes a technician is the highest trained technical person in the company, while in other situations technicians work for a research engineer. In the past, technicians would locate and replace discrete components of a system, but the troubleshooting of the future is at board level and requires a more systemic approach.

A Workforce Example

Although specific to the Semiconductor industry, the Maricopa Advanced Technology Education Center (MATEC) has provided a clear definition of what SEMATECH and SEMI/SEMATECH consider as a Semiconductor Equipment Technician's job. This definition provides generic guidelines for the skills required for *any electronics oriented technician* in today's job market.

Required Education and Experience: Associate degree in electronics, semiconductor manufacturing, microelectronics *or* related technical field or equivalent experience.

General Job Duties: Monitor, maintain and perform a variety of complex repairs on semiconductor wafer fabrication equipment to ensure uninterrupted production flow. Perform periodic preventive maintenance procedures as defined by specifications.

Provide technical support in the form of troubleshooting, installation, diagnostics, adjustment, repair, modification, assembly and calibration of equipment according to specifications, blueprints, manuals, drawings and verbal or written instructions. Use a structured and comprehensive method to identify the root cause of process or equipment malfunction; implement corrective action after thorough analysis to increase probability of the right fix the first time, based on product quality parameters.

Perform electrical, mechanical software troubleshooting and maintenance for related equipment tools, cable assemblies and fixtures. Check and calibrate tools, equipment and fixtures using test and diagnostic equipment as required. Clean and lubricate shafts, bearings, gears and other parts of machinery.

Assist in the layout, assembly, installation and maintenance of pipe systems and related equipment. Maintain and monitor maintenance parts stack. Maintain accurate records and logs of work performed, modifications, calibrations, and adjustments and parts inventory.

May perform equipment and fixture modifications as directed by manufacturing engineers. Equipment used includes office equipment, power supplies, oscilloscopes, logic analyzers, volt meters, soldering irons, hand tools, power tools and personal computers or other hand/power tools and test equipment. Maintain proficiency in programmable controllers, microprocessors, control circuits, analog/digital circuits, motors and troubleshooting skills.

Understanding and Content

The most important general goals for the mathematical education of students in technical electronics fields are

- development of formula manipulation skills and understanding of the value of the formulas via the use of real problems
- development of problem solving skills and critical reasoning.

Mathematical Problem Solving Skills. Students coming into technical areas need more training in problem solving: the identification and description of problems, the translation of English prose into well-formed mathematical equations, and the interpretation of mathematical solutions in terms of physical reality.

Students must learn how to set up a problem using appropriate units and satisfying the requirements of dimensional analysis. They must learn how to identify relevant information from that which is superfluous or extraneous.

Students must be able to apply logical analysis to solve multi-step problems: identify the individual steps and concepts necessary to reach a solution, then use appropriate mathematical tools to execute the plan. At the most basic level students must know which (theoretical) equations are needed to solve a particular problem and then be able to substitute known values into these equations and solve for the desired unknown.

Students must learn to interpret their mathematical solutions in the context of the physical problem with which they began. They should learn how to perform an error analysis. They must also learn to estimate answers and to obtain approximate solutions: this provides an important tool for detecting mistakes and inaccurate answers. In general, to become good problem solvers students must develop a sense of careful time-management, good judgment, and tenacity.

Specific Mathematical Topics. The following are topics that are important for students planning on careers in technical fields of electronics.

Basic Topics

- The metric system
  - Measurements and units
  - Conversions between and within system
  - Dimensional analysis
  - Consistent use of units
- Concepts of perimeter, area, volume
  - "How many atoms are in this cell?"
  - Visualization of inverse-square phenomena
- Estimating and using orders of magnitude

Algebraic Topics and Skills

- Solving equations for particular variables
- Linear equations and slopes
- Quadratic equations
  - Important to test solutions for feasibility.
  - Needed for complex numbers and conjugates
  - Necessary for considering nonlinear effects
- Simultaneous equations (usually two equations)
  - Cramer's rule, Gaussian elimination
  - Matrix methods (for more than two variables)
- Manipulation of inequalities
- Number systems: binary, hexadecimal

Functions

- Definitions of function, dependent variable, and independent variable
- Absolute value functions, step functions, polynomials, linear functions & slopes
- Exponential functions and logarithms (base 10)
- Composite functions
- Limits of functions, zeroes, asymptotic behavior, extrapolation
  - Semiconductor example: where does Ohm's law break down?

Trigonometry

- Basic trigonometric functions: sine, cosine, and tangent
  - Computations with right triangles
  - Relationships with the unit circle
- Only the *most basic* trigonometric identities should be introduced

Radian measure

Graphical analysis of sines and cosines:

Amplitude, phase, frequency, RMS peak, and relationship between RMS and peak.

Vectors

Resultants, addition and subtraction of vectors

Vector products and cross-products (for higher level degrees)

Phasors (vectors in the complex plane representing sinusoidal signals)

Graphs of Functions and Data

Coordinate systems, both rectangular and polar

Use of log and semi-log paper

Curve-fitting techniques (linear regression, etc.)

Graphical analysis

Statistics

Means, medians, standard deviations

Normal distributions

Sigma notation

Know when a process is "drifting" (statistical controls)

Variability in measurements

Difference between population and sample (discrete vs. continuous data)

Descriptive statistics (skew)

Quality control

Differential and Integral Calculus

(This is most appropriate for engineering technicians.)

Difference quotients ("differencing") as related to velocity, acceleration, etc.

Derivatives and integrals and their meanings

Delta functions

Technology

There should be a balance between students' utilization of technology versus manual computation and reasoning from theoretical knowledge.

Relying solely on computer-based instruction or video-based instruction is not effective. Use of computer support is more effective. Instructors should demonstrate computational techniques and software during class sessions.

Web assisted or enhanced courses can be very effective. Instructors should consider developing course web sites, including chat groups. Establish e-mail communication and bulletin boards: encouraging students to seek instructional assistance in this manner can be very effective.

Future employers are likely to each have their own preferred computer computation software such as Mathematica, Maple, MathCad, or MathLab. Students should ideally develop proficiency with at least one such program, as well as a working knowledge of spreadsheets. It is also useful for students to develop three-dimensional visualization skills via programs such as AutoCAD.

Instructors should not promote over-reliance on hand-held calculators. In particular, teaching with calculators alone might be a disservice if students ultimately meet only computers in the workplace. If instruction is centered on calculators, then it must be done in such a way that students will be able to carry the knowledge they gain from calculators over to computer applications.

Instructional Techniques

Mathematics instructors should ensure that active learning occurs in their classes. Require students to think on their feet, to present demonstrations in class, and in general to take responsibility for their own education. There should be an emphasis on communicating mathematics, both in written and verbal form; in particular, students should be required to produce written and oral reports. Students need to see instructor derivations in order to appreciate the theory, but they must have classroom activities that develop their own skills at working with the material.

Instructors should accommodate different learning styles. This can be accomplished by using a variety of class session formats: lectures, demonstrations, class participation, and experiments. Lectures delivered in small bursts are more effective than long monologues. Employ drill and practice when necessary: many algorithms and calculational techniques require this. Encourage note-taking: it is beneficial for students to regurgitate what's been said in class. It helps them prioritize information and rethink what they have learned.

Use assignments built around multi-step projects. Such projects are often ideal for collaborative team efforts. Learning how to work effectively in a team is a skill that cannot be overemphasized for future employment. Students should further be encouraged to form their own study groups.

Instructional Interconnections

There needs to be more cooperation and exchange of information and instructional materials between faculty members in mathematics and faculty members in the technical fields. The individual departments are often too isolated and insular. Arrange for interdepartmental visits. Involve faculty members from different departments on advisory boards or in hiring decisions.

Encourage faculty members in the technical fields, as well as faculty in physics and chemistry, to supply meaningful problems to the mathematics instructors. Mathematics instructors should be encouraged to attend the technical courses to gather further useful examples. Everyone wins — students and instructors — because concepts from the various technical and scientific disciplines are then reinforced in the mathematics courses.

Look for ways to more effectively integrate the content of mathematics courses with the other disciplines. In addition to the incorporation of realistic examples, consider having appropriate courses team-taught by faculty from both mathematics and the technical fields. Reach beyond the college or university: invite industry representatives to make presentations in the classroom.

WORKSHOP PARTICIPANTS

Gregg Anderson, Materials Engineer, Air Force Research Lab (AFRL/MLPJ) USAF, Wright Patterson Air Force Base, Dayton, Ohio.

David Baker, Professor Emeritus, Rochester Institute of Technology/Past Chair of TAC of ABET.

Elizabeth Basinet, President, Barrett Resource Group, San Diego, California.

Bob L. Bixler, Workforce Consulting, Austin Community College, Austin, Texas (Currently at San Juan Basin Technical School, Cortez, Colorado).

Steve Cooperman, Science Teacher Lecturer, Westridge School (high school), Pasadena, California.

Christopher Dennis, Consultant, Steaming Kettle Consulting, Portland Community, Portland, Oregon College.

Linda Hollstegge, Instructor, Information & Engineering Technology, Cincinnati State Technical & Community College, Cincinnati, Ohio.

James Hyder, F11X Training Developer (Thin Films), Intel Corporation, Hillsboro, Oregon.

Gerald Janey, Chairman of Engineering Technology, Massasoit Community College, Brockton, Massachusetts.

Rino Mazzucco, Instructor, Mesa Community College, Mesa, Arizona.

Lyle B. McCurdy, Professor Electrical & Computer Engineering Technology, Dept. of Engineering Technology, California State Polytechnic University, Pomona, California.

John Palmer, Professor of Manufacturing Engineering/Electronics Technology Program, Industrial/Electronics, Central Arizona College, Signal Peak Campus, Coolidge, Arizona.

James Payne, Supervisor Tech Support & Management Systems, Oak Ridge National Laboratory, Oak Ridge, Tennessee.

Branisalv Rosul, Professor, College of DuPage, Glen Ellyn, Illinois.

Tim Woo, Instructor, Electronics/Engineering, Durham Technical Community College, Durham, North Carolina.

Jim Wood, Division Chair, Industrial & Engineering Technology, Tricounty Technical College, Pendleton, South Carolina.

ACKNOWLEDGEMENTS

The members of the panels would like to thank Bruce Yoshiwara, of Los Angeles Pierce Community College and Gwen Turbeville, of J. Sargeant Reynolds Community College, for coordinating the workshops at their respective campuses. We would also like to thank the Mathematical Association of America for initiating this series of workshops and to The American Mathematical Association of Two-Year Colleges for providing this opportunity to consider and make recommendations on a topic of importance to the information technology community, especially project directors Mary Ann Hovis, Rob L. Kimball, and John C. Peterson. Finally, we would like to thank the National Science Foundation for providing the funds to support these meetings.

APPENDIX A: Definitions

MATEC is the Maricopa Advanced Technology Education Center, a Division of Maricopa Community Colleges, Tempe, Arizona.

SEMATECH is Semiconductor Manufacturing Technology. International SEMATECH is a unique endeavor of 12 semiconductor-manufacturing companies from seven countries. Member companies include AMD, Agere Systems, Hewlett-Packard, Hynix, Infineon Technologies, IBM, Intel, Motorola, Philips, STMicroelectronics, TSMC, and Texas Instruments. For more information see `www.sematech.org`.

SEMI/SEMATECH is a consortium of U.S. semiconductor suppliers. Here SEMI is an acronym for Semiconductor Equipment and Materials International. SEMI/SEMATECH and SEMATECH, the sister consortium of U.S. semiconductor manufacturers, were founded in 1987. SEMI/SEMATECH members are suppliers to all areas of semiconductor manufacturing. SEMI/SEMATECH is the primary interface between its members and SEMATECH and International SEMATECH. The purpose of SEMI/SEMATECH is to ensure the health of the U.S. semiconductor supplier infrastructure. As the voice of the U.S. semiconductor supplier infrastructure, SEMI/SEMATECH fosters communication between its members and their customers within and beyond SEMATECH.

Skill standards are the quality standards applied to people. They are specific statements of desired skills and knowledge presented in an observable and measurable form. A standard consists of a condition, a behavior and the standard to which the behavior should be performed.

MATEC has created a document called *Semiconductor Manufacturing Technician Skill Standards*, which can be purchased at `www.matec.org`. This document clearly defines all of the skill standards that have been identified for technicians in the semiconductor industry ranked by importance, proficiency, frequency and difficulty.

Tasks: While it is important to catalogue skills, it is not always possible to determine the skills requisite to each functional area within the semiconductor-manufacturing environment until the tasks required at each level have been ascertained and analyzed. A task, for the purpose of this definition, is an activity or procedure assigned as part of a technician's job in the factory.

Training Levels: Groups of tasks that define the degree of proficiency for a technician's job performance. SEMATECH and SEMI/SEMATECH recognize three general levels of proficiency. These are:

Level 1: Tasks are a **combination of basic tasks** in the process, operations and equipment categories. *The tasks involve limited decision-making and are frequently performed.*

Level 2: Tasks are **intermediate tasks** in each category. *The tasks require decision-making as prescribed in controlled documents (specifications and response flow charts) and are frequently performed.*

Level 3: Tasks consist of **advanced tasks** in each category. *The tasks require decision-making based on individual judgment and are infrequently performed.*

APPENDIX B: Examples and Vignettes

Electronic Examples

These are sample problems from electronics. Not all programs include all of these problems, and not all types of problems are illustrated. This list does not prescribe what teachers of electronics expect math departments to teach; it illustrates applications in electronics.

1. Convert 26 milliamperes to microamperes and to amperes.

   Answer: 26 milliamperes is 26×10^{-3} amperes and is 0.026×10^{-6} amperes which is 0.026 μA or 0.026 microamperes.

2. Find the total resistance of three resistors in parallel. The formula is $R_t = 1/(1/R_1 + 1/R_2 + 1/R_3)$ where $R_1 = 150\ \Omega$, $R_2 = 300\ \Omega$, $R_3 = 900\ \Omega$

   Answer: 90 Ω

3. Find the resonant frequency of a parallel RC circuit containing a capacitor of 20μfd and an inductor of 30 μH. The formula is $1/(2\pi[LC]^{1/2})$

   Answer: 6497 Hz

4. A series RLC circuit contains 100Ω inductive reactance, 60Ω capacitive reactance, and 30Ω resistance. Find the polar resultant. The inductance is a vector up while the capacitance is a vector down and the resistance is to the right. The inductance and capacitance are on the imaginary axis while the resistance is on the real axis.

   Answer: The solution is a vector 50Ω in magnitude at a positive angle of 53.1 degrees as shown in Figure 1.

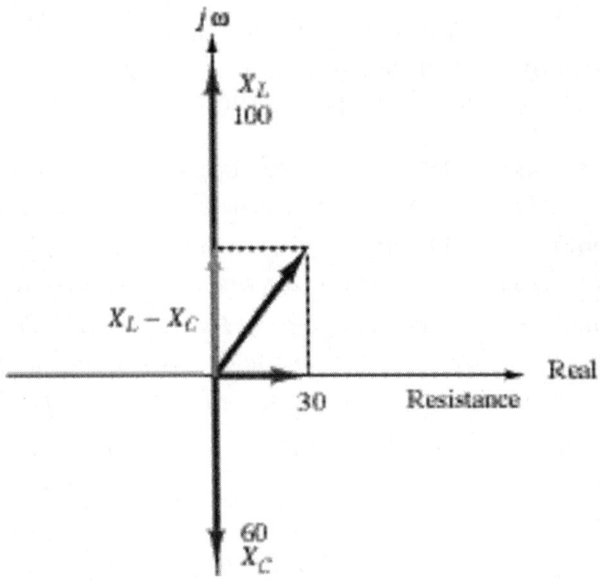

Figure 1

5. A capacitor of 0.01μF is connected in series with a resistor of 100 KΩ and a battery of 100 volts and a switch as in Figure 2. Graph the voltage across the capacitor in response to closing the switch. The

response is described by the equation:

$$V_c = V_F + (V_o - V_F)\varepsilon^{-t/\tau}$$

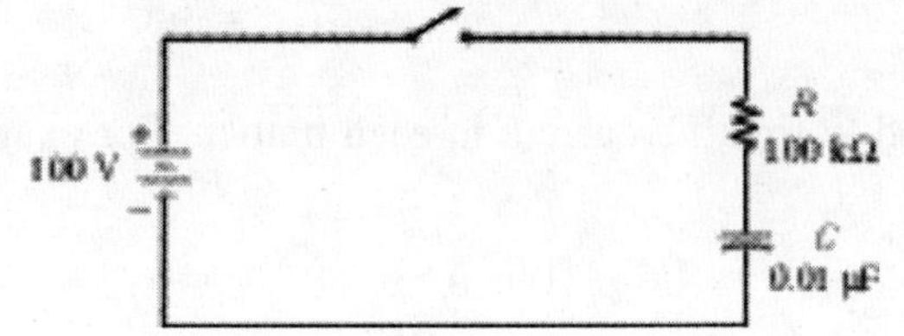

Figure 2

Solution: The full charge voltage is 100 V, which is the 100% level (1.0) on the graph. 75V is 75% of the maximum, or 0.75 on the graph in Figure 3. You can see that this value occurs at 1.4 time constants. One time constant is 1 ms. Therefore the capacitor voltage reaches 75 V at 1.4 ms after the switch is closed.

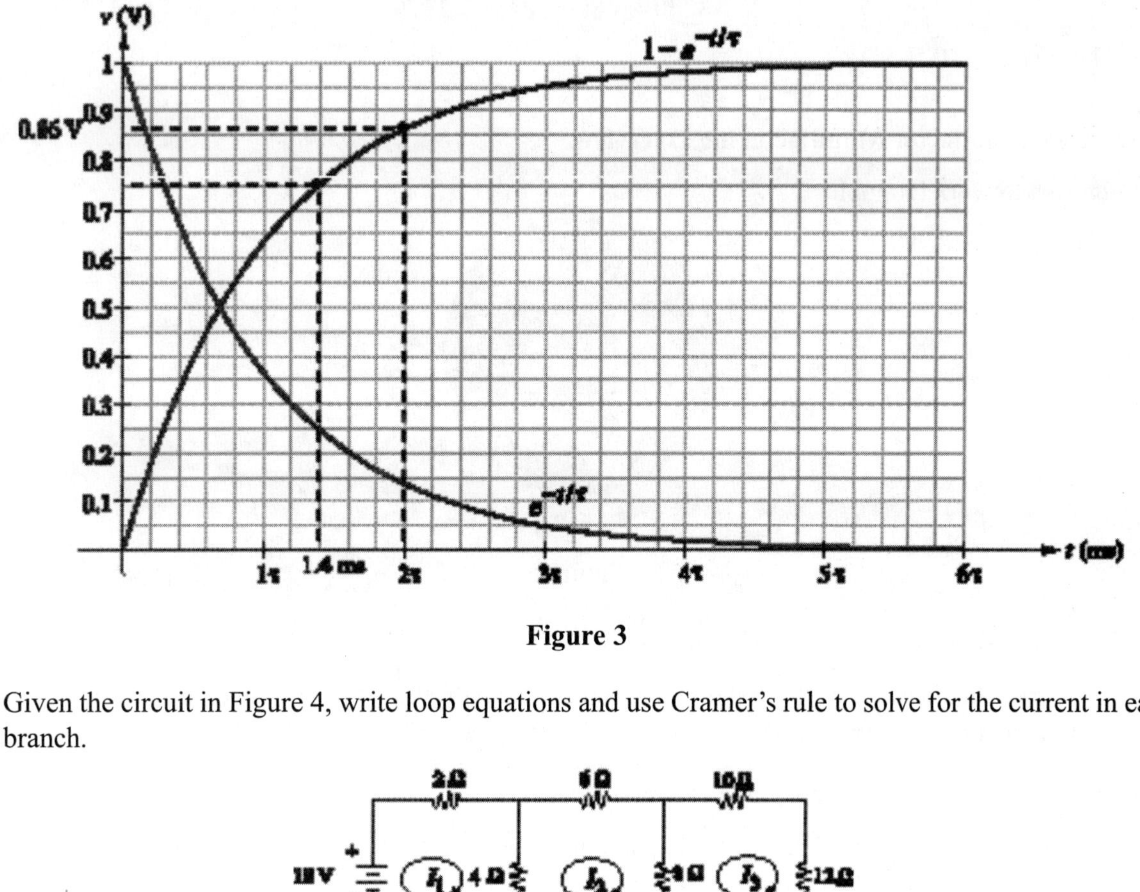

Figure 3

6. Given the circuit in Figure 4, write loop equations and use Cramer's rule to solve for the current in each branch.

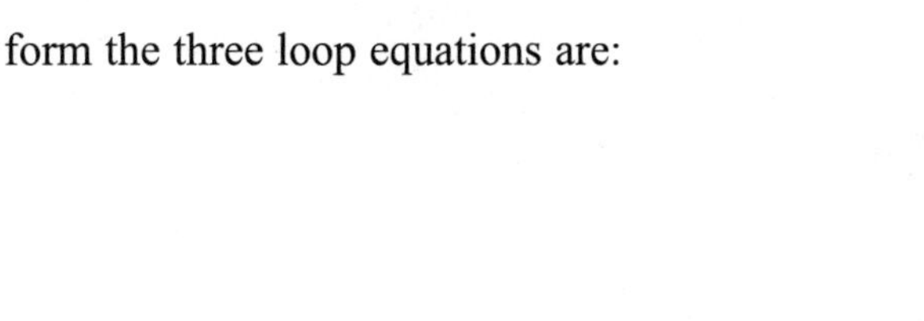

Figure 4

Solution: In standard form the three loop equations are:

$$
\begin{array}{rcrcrcl}
-6I_1 & - & 4I_2 & & & = & 10 \\
-4I_1 & + & 18I_2 & - & 8I_3 & = & 0 \\
 & - & 8I_2 & + & 30I_3 & = & 0
\end{array}
$$

Cramer's rule can be used to find the current in each branch. For example, using Cramer's rule, the current I_2 is given by

$$
I_2 = \frac{\begin{vmatrix} 6 & 10 & 0 \\ -4 & 0 & -8 \\ 0 & 0 & 30 \end{vmatrix}}{\begin{vmatrix} 6 & -4 & 0 \\ -4 & 18 & -8 \\ 0 & -8 & 30 \end{vmatrix}}
$$

$$
= \frac{0-(-1200)}{3240-(480+384)^2} = \frac{1200}{2376}
$$

Therefore, $I_2 = 0.505$ A.

General Semiconductor Manufacturing Overview

Use Figure 5 to work Examples 7–9.

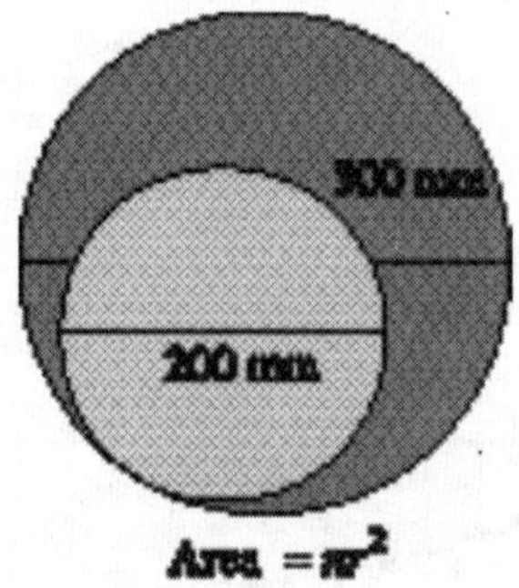

Figure 5

7. What is the manufacturable area of a 200 mm wafer?

   Solution: The diameter of the wafer is 200 mm, so its radius is 100 mm. This means that the manufacturable area is $\pi r^2 \approx 3.14(100 \times 100) = 31{,}400$ square mm.

8. What is the manufacturable area on a 300 mm wafer?

   Solution: The diameter of this wafer is 300 mm, so its radius is 150 mm. Thus, the manufacturable area is $\pi r^2 \approx 3.14(150 \times 150) = 70{,}650$ square mm.

9. How much more manufacturability does a 300mm wafer provide over a 200mm wafer?

   Solution: There are two ways to respond to this question.

   Ratio method: The ratio of the two areas is $\frac{70{,}650 \text{ mm}^2}{31{,}400 \text{ mm}^2} = 2.25$. This means that a 300 mm wafer has an area that is 2.25 times the area of a 200 mm wafer.

Difference method: This method just subtracts the two areas.

$$70{,}650 \text{ mm}^2 - 31{,}400 \text{ mm}^2 = 39{,}250 \text{ mm}^2.$$

The area of a 300 mm wafer is 39,250 mm^2 more than the area of a 200 mm wafer.

An area in the Fab has eight tools, which process 11 different layers for the factory. Each time a wafer passes through the area it counts as one wafer. The area can produce 65,000 wafers a week; at 100% tool availability and 100% tool utilization. **Tool availability** is the measure of how much time during the week that the tool is available for running wafers. **Tool utilization** is the measure of how much time during the week that the tool is working on product. Knowing this please answer the Exercises 10–14.

10. How many wafers will the area have to run on each tool in order to maintain 65,000 wafers a week?

    Answer: 65000/8 = 8125

11. Assuming 100% tool availability and 100% tool utilization, what is the processing time for one lot? (Assume 25 wafers per lot).

    Answer:

$$\begin{aligned} 65000/25 &= 2600 && \text{lots/week per area} \\ 2600/7 &\approx 371.43 && \text{lots/day per area} \\ 371.43/8 &\approx 46.43 && \text{lots/day per tool} \\ 46.43/24 &\approx 1.935 && \text{lots/hour per tool} \\ 60/1.935 &\approx 31 && \text{minutes/lot} \end{aligned}$$

    It will take about 31 minutes to process one lot.

12. Assuming 80% tool availability and 100% tool utilization, how many wafers can the area produce?

    Answer: Using the lot run time of 31 minutes, with the tool available only 80% of the time during the week, we get the following:

$$\begin{aligned} 60\times 24\times 7 &= 10080 && \text{minutes in the week} \\ 10080\times 0.8 &= 8064 && \text{adjusted time for processing wafers} \\ \frac{8064}{31} &\approx 260.13 && \text{lots possible to produce on one tool} \\ 260.13\times 8 &= 2081.04 && \text{lots possible to produce in area} \\ 2081\times 25 &= 52025 && \text{wafers possible to produce in area} \end{aligned}$$

    Thus, with 80% tool availability and 100% tool utilization, it is possible to produce about 52,025 wafers in a week (allowing for rounding errors).

13. Assuming 80% tool availability and 80% tool utilization, how many wafers can the area produce?

    Answer: $80\%\times 80\%\times\left(\dfrac{60*24*7}{31}\right)\times 8\times 25 = 41620$

    The area can produce about 41,620 wafers.

14. What would the lot run time have to be in order to meet the goal of 65,000 wafers a week?

Answer: $\dfrac{0.8\times0.8\times60\times24\times7\times8\times25}{65{,}000}\approx 19.85$. It will take about 19.85 minutes per lot.

Sample Etch Math Problems

15. When you measure oxide on a wafer before etch, it is 5000 Angstroms thick. After two minutes of etching, it has 2000. Etch rate is defined as the rate by which material is removed from a wafer. What is the etch rate?

Answer:

$$\text{Incoming Thickness} - \text{Outgoing Thickness} = \text{Amount Etched}$$

$$\frac{\text{Amount Etched}}{\text{Etch Time}} = \text{Etch Rate}$$

The amount etched is

$$5000\text{A} - 2000\text{A} = 3000\text{A}$$

$$\frac{3000\text{A}}{2\text{ min}} = 1500\text{ A/min}$$

The etch rate is 1500A per minute.

16. According to your engineer, the etch rate in Problem 15 should be 1.5×10^2 per minute.
(a) Does the etch rate that you measured match this number?
(b) How much faster or slower is the above etch rate?
(c) Why is this important?

Solution:

The etch rate should be $1.5 \times 10^2 = 150$ A per minute. In Problem 15 we found that the etch rate was 1500 A per minute. This does not match the engineer's number.

It is 10 times faster than it should be.

These wafers would be scrapped due to over-etching the wafers.

17. A gas flow of 350 ccm can vary by 5% and not disrupt the process. What is the highest the gas flow can be? The lowest?

Solution:

$$350 \times 0.05 = 13.5$$

$$350 + 13.5\text{ ccm} = 367.5\text{ ccm}$$

$$350 - 13.5\text{ ccm} = 337.5\text{ ccm}$$

Highest = 367.5ccm, Lowest = 337.5ccm.

Sample Thin Films Problem

The etched patterns on a wafer now are called vias. As an insulator or a metal is deposited on a wafer, all patterns that have been etched into the surface are filled (as illustrated in Figure 6).

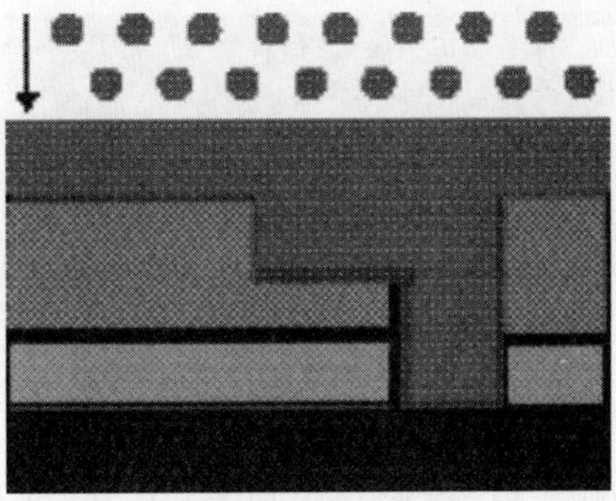

Figure 6

Determine what to set the **Dep Time (DT)** to in order to bring wafer thickness back into specification. Dep Time is the amount of time that an insulator or a metal layer is being deposited on the wafer.

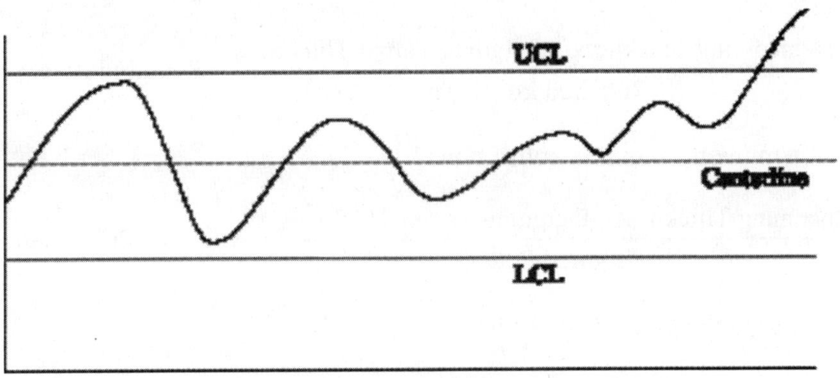

Figure 7

The Upper Control Limit (UCL) is 500 angstroms (see Figure 7) and the Lower Control Limit (LCL) is 400 angstroms. The last test fire was 525 Angstroms. Current Dep Time for the tool is 4.28 minutes. We generally target the tool to centerline to provide for the greatest amount of variability within the process. The Dep Time to Thickness ratio is generally considered to be a linear relationship.

18. What Dep Time will bring the tool back to the centerline thickness?

Answer:

$$\frac{525 \text{ Angstroms}}{4.28 \text{ min}} = \frac{450 \text{ Angstroms}}{x}$$

$$x = \frac{450\text{A} \cdot 4.28 \text{ min}}{525\text{A}}$$

$$x \approx 3.66 \text{ min}$$

A Dep Time of about 3.66 minutes will bring the tool back to the centerline thickness.

Sample Planarization Problem

After all of the vias in the wafer are filled with a layer, the surface must be planarized (or since it is smoothed it is also called polished) as illustrated in Figure 8.

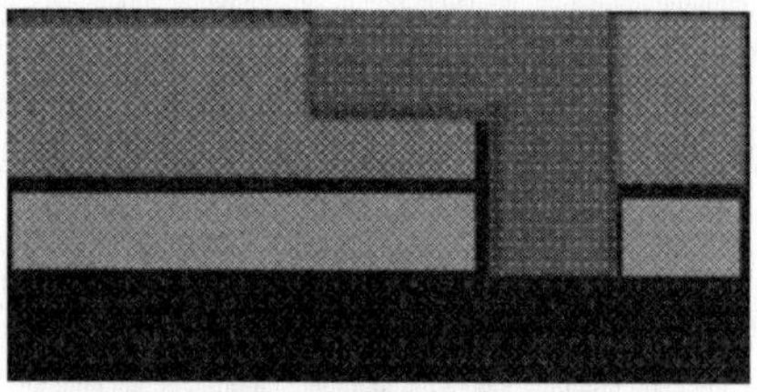

Figure 8

To create such a uniform surface, a precise incoming thickness and outgoing target thickness is required. The tool generally calculates the polish rate automatically.

19. Assuming a planarization rate of 1000 Angstroms/minute, how much time is required on the tool if the incoming thickness is 5000 Angstroms and the desired outgoing target is 3500 Angstroms?

Solution: First we determine the polish rate:

$$\frac{\text{Incoming Thickness} - \text{Outgoing Target Thickness}}{\text{Required Polish Time}} = \text{Required Rate}$$

Next we determine the required polish time:

$$\frac{\text{Incoming Thickness} - \text{Outgoing Target Thickness}}{\text{Required Rate}} = \text{Required Polish Time}$$

Applying these to the given data, we have:

$$\begin{aligned}\frac{5000\text{ A} - 3500\text{A}}{1000\text{ Angstroms per minute}} &= \frac{1500\text{A}}{1000\text{ A/min}}\\ &= 1.5\text{ min}\end{aligned}$$

So, 1.5 minutes is required to hit the target thickness.

Technical Mathematics: Information Technology[1]

CRAFTY Curriculum Foundations Project
Los Angeles Pierce College, October 5–8, 2000
J. Sargeant Reynolds Community College, October 12–15, 2000

Robert D. Campbell, John C. Peterson, and Kathy Yoshiwara, Report Editors
Bruce Yoshiwara and Gwen Turbeville, Workshop Organizers

Summary

Information Technology (IT) is a relatively young and rapidly evolving field. This ever-changing environment makes it difficult to identify the specific mathematical content in job skills for individual IT positions; in fact, many IT jobs require few particular mathematical skills. Significant developments in the IT realm, including web-based environments, suggest a focus on nonlinear thought processes, mathematical reasoning skills, and creative problem solving, rather than specific content. Therefore, academic mathematics preparation for students pursuing IT careers should not require advanced topics but should instead provide a solid foundation of elementary content, with a strong emphasis on the analytical ability needed to understand mathematical concepts.

These fundamental concepts and skills must prepare a student to enter the field initially, and must also provide a basis for lifelong learning, which may include seeking additional degrees. IT workers rarely need additional mathematical skills for professional development and career advancement, but do draw on the analytical skills acquired during their initial mathematics training.

Throughout this report, we emphasize that content and pedagogy should connect theory with applications. While few technicians use advanced mathematics on a daily basis, the ability to relate the correct mathematical concept to the problem at hand creates an IT technician with a future. The transfer of foundational knowledge to application skills early in the education of IT students will serve them well and will help them transfer these same concepts to new technologies in the course of their evolving careers. A technician educated in this manner is an asset to his or her organization and has the training needed to progress up the career ladder.

Many two-year schools must teach mathematical topics required by four-year colleges pursuant to transfer arrangements and articulation agreements. These agreements are often designed to enable students to continue to a baccalaureate degree, even though a high percentage of two-year students have no desire to follow that path. Although this mismatch was raised as a concern, it was deemed outside the scope of our discussion. For the purpose of this report, we focus on the mathematical skills students should master while completing the associate degree as an entry into the job market.

[1] This material is based on work supported by the National Science Foundation under DUE grant no. 0003065. Any opinions, findings, and conclusions or recommendations expressed in this paper are those of the authors and do not necessarily reflect the views of the National Science Foundation.

Narrative

Introduction and Background

Technology changes very rapidly. “Timeless” skills that are relevant today and will remain relevant in the future are therefore very desirable. Technology demands problem-solving skills, a range of analysis tools from simple to sophisticated, the ability to identify and probe various approaches to a problem, and the ability to synthesize information to reach meaningful conclusions. These skills are useful in other disciplines as well as in the job environment. For example, students must be able to trouble-shoot (solve problems) in computer classes.

Students should be introduced to newer and more sophisticated technological tools used within their trade as they are developed. However, these tools should never replace abilities such as estimating, performing simple mental arithmetic with precision, or evaluating a tool’s accuracy. Students should develop a comfort level with a variety of tools, an understanding of the associated applications, and a sound grasp of the mathematical concepts associated with the applications. They must understand the concepts in order to analyze a problem and select the most appropriate, efficient, and effective tool(s) to solve it.

Technology has eliminated the need for students to concentrate on the mechanics of mathematics, and attention has shifted to mastering tools. Tools as simple as spreadsheets and calculators are replacing the need for proficient computation. For example, statistical quantities are rarely computed today without such technological tools. To the extent that mathematics courses teach and reinforce the use of technological tools, IT students are well served.

Technology spans geographical and cultural differences today more than ever before. Students must learn to work effectively and efficiently no matter where their jobs take them. They must be comfortable working in various systems and moving between them in our global society. Currency conversion and fluency with the metric system of measurement are two simple examples of skills they will need in order to succeed. Furthermore, technology demands the ability to work in teams and to collaborate with others, and can thus influence the pedagogy of various disciplines.

Web-based environments and other IT settings suggest the need for nonlinear thought processes, and for one-to-many or many-to-many relationships, outside the traditional function model. Topics such as fractals, combinatorics and graph theory support this need, as do activities that require creative problem solving. For example, a computer security company may hire employees who are good hackers: people who are oriented towards nonlinear logic, people who find alternate paths. Analytical skills are still needed, but reliance on specific mathematical skills is diminishing.

Understanding and Content

We identify below the specific fundamental mathematical content appropriate for students pursuing IT technician careers through community college credentials. Although the particular mathematical concepts and skills needed may vary depending on individual IT career tracks, we believe all IT technicians would be well served by mastering this collection. It is also important to note that although not all of these concepts and skills are used on a day-to-day basis in targeted IT jobs, the mathematical environment supports the analytical mental training necessary for success and advancement in the IT field.

Foundation Content. Specific foundation content includes:

Basic arithmetic skills, including computational skills and use of calculators, decimal arithmetic, mental arithmetic, fractions, percentages, approximation, truncation and rounding, working with formulas and problems without “nice” answers.

Estimation skills and the ability to determine the reasonableness of an answer. (Examples of applications include estimating the volume of data in large databases, judging whether a report or calculation gives an answer of the correct magnitude, and determining whether a system is operating within expected ranges of performance.)

Conversions between different measurement systems, and knowledge of the metric system. (Accounting systems in the global market, which work with currency other than dollars, are an example of an application.)

Working with different bases (decimal, binary, octal, hexadecimal), numerical operations within the base, conversions between bases (including rounding errors, and divisibility issues). Applications include calculation of IP subnet masking required to support the number of subnets and hosts per subnet, and rounding or truncation errors caused by converting from the binary number system to base 10.

Basic geometry concepts, including perimeter, area, volume. (Design and deployment of a security camera system is an example of an application.)

Boolean algebra concepts, Boolean values and fundamental operations on Boolean values. (Expression of electrical networks in Boolean notation as an aid in the development of switching theory and in computer design is an example of an application.)

Fundamental concepts and skills of algebra, including variable manipulation and solving for a variable, linear systems of multiple variables, graphing in two dimensions, definition and basic properties of functions, basic properties of matrices, algorithms. (Spreadsheets and tables are applications of matrices.)

Fundamentals of statistics and probability, including data analysis and presentation, descriptive statistics, use of spreadsheets, and the use of probability and statistical models to draw inferences. (The collection, monitoring and interpretation of network/system performance measurements for network traffic control and load balancing are common job responsibilities for IT technicians. Other applications include large databases where reports are meaningful only when data are filtered through statistical terms, and the notion of "five 9s" of reliability, which refers to 99.999% uptime for systems such as telephones that must be dependable for 911 calls.)

Basic right triangle trigonometry, elementary trigonometric functions and their graphs. (Wavelength processing and interference issues in cabling systems are examples of applications.)

Accounting and related business mathematics concepts and terminology. (The wide range of IT applications dealing with financial systems provide examples. Although accountants may direct decisions, a basic understanding of terminology and concepts is very valuable for technicians. Other applications include cost/benefit analysis, ROI (return on investment), and payback concepts for interpreting technology decision support packages.)

Fundamentals of logic, logical connectives, truth tables, deductive reasoning, digital logic, logic gates, flip-flops. (The logic within the if… then/and… or statements inherent in software applications is an application, as are the basic troubleshooting approach to a systems failure for a system with *n* components, and the design of a sample data set to test a newly developed software system.)

While an in-depth exposure to mathematical theory may not be necessary, we believe that a survey course of miscellaneous topics emphasizing a variety of mathematical models is valuable. Mathematical modeling, problem-solving techniques, and challenging activities would be beneficial and appropriate. Fairly traditional content might serve the purpose, but it should incorporate nontraditional topics such as fractals, combinatorics, and graph theory that support creative problem solving. Students would be well served by skills in algorithm development and pattern analysis, as well as exposure to relational algebra, queuing theory, and set theory. An emphasis on multiple representations of data and visualization is very important.

Content for Advancement. Students may need additional mathematics for advancement up the career ladder or continuation into a baccalaureate program. It may be necessary for students to complete a bridge course in order to enter a baccalaureate program; such a course would expand upon the discrete mathematics concepts listed above and provide additional depth in such topics as set theory, counting theory, proofs, sequences and series, analysis of functions, and algorithmic design and analysis. Although we have not identified calculus as necessary or appropriate for IT, many computing degree programs do require calculus. Therefore IT students may need to study calculus in order to pursue baccalaureate or advanced degrees.

Mathematical Problem Solving Skills. All IT students should become expert problem solvers. To achieve this goal they do not need an extensive knowledge of algebra, but they do need the logical problem solving process used in solving algebraic problems. Algebraic problem solving skills include writing an equation to represent a problem: the dreaded "word problem." IT students must be able to analyze a situation and develop an equation before the ability to solve an equation of a particular type is of value.

The ability to transfer information from one setting to another is very important. For example, even though an individual may be able to solve an equation, he or she must be able to transfer that ability to solving real-world problems. Problems that require students to apply a skill or concept from one area to a wholly different area are very valuable learning experiences.

Students should use a variety of problem-solving strategies (such as divide and conquer) to solve IT-based problems. Integrating mathematical concepts with practical application skills taken from the IT industry provides students with a reference point for the mathematics being studied. Application-based problems help students learn to derive equations from data and then conduct the appropriate analysis. Students should define the problem, collect the relevant data, perform all analysis, and make final recommendations in a report format.

In the field of IT, memorization is less important than the ability to use reference materials and other resources effectively. Problem-solving and analytic skills are critically important because, in the information age and in the IT field, there is simply too much to know. We recommend moving away from computational activities to more abstract problem solving and brainteaser exercises. Instructional techniques should emphasize the growing importance of creative team problem solving.

The problem-solving skills discussed in this report provide the mental processes needed to solve problems within IT systems. It is not necessary that students fully understand the related concepts; they should know they exist, how they are used, and how to choose the appropriate tool for a problem.

Technology

New technologies provide a multitude of instructional techniques. This rich environment can include both synchronous and asynchronous activities, providing students with opportunities to learn without being bound to a particular location.

Multimedia delivery systems, coupled with asynchronous personal contact time between students and faculty, allow instructors to cover difficult and challenging topics in more targeted ways. The asynchronous component of a course using this format should provide students with deeper understanding of the difficult concepts. As these new technologies are deployed, efforts to address a variety of learning styles and to provide all students with environments that promote individualized learning should be emphasized.

As instructional methods emerge, are tested and evolve, identifying methods that deliver the most effective and efficient learning experiences should be a priority. This emphasis on both quality of instruction and quantity of time invested reflects a growing awareness of cost-benefit analysis for education and its subsequent application in the workplace.

Instructional Techniques

IT-based applications should drive the development of mathematical theory and its use. At present it appears that theory is taught first, followed by application skills. In many cases, a mathematics professor teaches the theory within a traditional mathematics course, and an IT professional teaches the application skills. Theory and applications should be interwoven and integrated. Or, at least, applications should be considered first, and then theory, to ensure that theory is related to real-world concepts.

Teaching mathematics as a laboratory course with applications relevant to IT would enhance comprehension for IT students. Such a laboratory course could be incorporated directly into the IT program if the mathematics material were presented in learning modules. These modules might provide "just-in-time" mathematics instruction, allowing students to discover mathematical concepts within their IT applications. By demonstrating the relationship between the concept and its use, just-in-time mathematics instruction would be very beneficial for IT students.

Bringing applications into the mathematics class is an alternative approach. Theory could be taught as a follow-up to solving practical problems, rather than using applications to supplement theoretical development. In such a setting, mathematics faculty and IT faculty should work together to select examples and case studies relevant to the mathematics. Mathematics and IT faculty could develop and team-teach courses for IT and its associated mathematics.

Each concept will need a different balance between theory and application to ensure understanding. However, we believe that for technicians this balance will usually be 40% theory and 60% application. This ratio is not a fixed standard but provides a guideline. The balance for a given topic will depend on several factors, including the IT career being served, the level of schooling, and the application being addressed. As a rule, instruction should be centered on technique and application but well grounded in theory.

Combining mathematical concepts with IT applications should create interest in the underlying mathematical theory. This interest will allow students to form conceptual relationships between theory and applications within their daily jobs, and they will be able to build upon their educational foundation to create life-long learning experiences.

The nature of IT involves the IT technician in a constant process of adding new information, techniques and abilities to his or her portfolio. The methods and schedule of course work delivered in the traditional model do not address this need. Modularized content with clear objectives and measurable outcomes and delivered in alternate formats is needed to address ongoing professional development for IT workers.

Instructional Interconnections

Participation in summer internships in industry is an excellent way for mathematics teachers to understand the practical aspects of mathematics in the IT field. Afterwards, teachers are better equipped to explain why specific skills are needed in specific jobs, and to connect mathematical concepts with IT applications.

Internships can be excellent experiences for both students and teachers, and are strongly recommended. Such experiences provide answers to (legitimate) student questions such as "When am I ever going to use this stuff?" An instructor's ability to seat skills in practical real-world settings is a great motivator for learning. The IT industry wants students to see specific examples of how topics in their mathematics classes will be used on the job.

Creating IT-based case studies and interdisciplinary scenarios in mathematics courses would ensure the integration of mathematical concepts and their application. These case studies and scenarios should include problem definition, data gathering, data analysis, and problem resolution. They will involve students in defining relevant mathematical concepts, identifying theories and tools needed to solve the problem, processing data, reporting results, and providing all relevant documentation. This program will also develop skills in the areas of teamwork, time management, organization, and interpersonal relationships.

To accomplish this interdisciplinary work, faculty from the IT department and the mathematics department must work as a team. The team should discover, discuss and implement the interweaving of course material. Administration, faculty, and the institution as a whole must be committed to this effort.

There is an emerging trend for students to enter the IT field directly out of high school with only industry certifications and no post-secondary education. In today's market, students see numerous job opportunities and sometimes view higher education as a time impediment to pursuing a career. That employers continue to seek candidates with certifications rather than degrees is of growing concern. We recommend that high school and college mathematics instructors work together to address the content recommendations of this report and to advise IT students that post-secondary coursework allows for the life-long learning necessary for long-term career advancement.

One might ask whether some mathematical reasoning skills could be obtained from other disciplines. For example, can the ability to quantify data, to graph and plot data, and to identify the significant numbers in that data be obtained from biology, chemistry and physics, or perhaps the social sciences? We believe that the answer can be yes. Therefore, we promote the notion of mathematics across the curriculum, so that mathematical applications and reasoning skills can be addressed and reinforced in a variety of settings.

Students also need a foundation in the historical development of computing, and this material is typically included in overview IT courses. However, treatment of the subject could be improved by a coordinated mathematics curriculum, and we recommend that institutions link mathematical content with topics being discussed simultaneously in IT courses. We regret that, in many settings, mathematics courses are intentionally viewed as filters for entry into IT curricula, and we strongly recommend a more collegial and collaborative approach to educating students.

REFERENCES

ACM Two-Year College Education Committee, *Guidelines For Associate-Degree Programs To Support Computing In A Networked Environment*. New York: Association for Computing Machinery, March 2000.

Freeman, Peter and William Aspray, *The Supply of Information Technology Workers in the United States*. Washington, DC: Computing Research Association, 1999.

WORKSHOP PARTICIPANTS

Connie Blackwood, Director of Education and Research Initiatives, Idaho Engineering Laboratory, Idaho Falls, Idaho.

Robert D. Campbell, Chief Information Officer & Executive Dean of Information Technology Services, Rock Valley College, Rockford, Illinois.

C. Fay Cover, Director, Division of Learning Technologies, Pikes Peak Community College, Colorado Springs, Colorado (Currently with Sun Microsystems.).

Robert Hillery, Chairman, Information Systems Technology, NHCTC Pease International Tradeport Campus, Portsmouth, New Hampshire.

Denecia Merritt-Damron, Information Technology Division Director and Instructor of Desktop Applications Software, Marshall Community and Technical College, Huntington, West Virginia.

Keith Morneau, Assistant Professor, Information Systems Technology, Northern Virginia Community College, Annandale, Virginia.

Scott Nicholas, MCSE and Instructor of Computer Networking, Marshall Community and Technical College, Huntington, West Virginia.

Randy Robertson, Network Instructor, Wake Technical Community College, Raleigh, North Carolina.

Dan Schwartz, Manager, Supply & Transportation Systems Development Group within IT, Marathon Ashland Petroleum, Findlay, Ohio.

Dick Thomas, Independent Database Consultant, Littleton, Colorado.

Nancy Thomas, Independent Database Consultant, Littleton, Colorado.

ACKNOWLEDGEMENTS

The members of the panels would like to thank Bruce Yoshiwara of Los Angeles Pierce Community College and Gwen Turbeville of J. Sargeant Reynolds Community College for coordinating the workshops at their respective campuses. We would also like to thank the Mathematical Association of America for initiating this series of workshops and to The American Mathematical Association of Two-Year Colleges for providing this opportunity to consider and make recommendations on a topic of importance to the information technology community; especially project directors Mary Ann Hovis, Rob L. Kimball, and John C. Peterson. Finally, we would like to thank the National Science Foundation for providing the funds to support these meetings.

APPENDIX A: Definitions

As we explore the mathematical skills needed by Information Technology technicians, some definitions and concepts must be provided. The following list creates the context for our report and should provide the reader with an understanding of our audience and the scope of our discussion.

Asynchronous methods: Non-simultaneous (and typically time-independent) activities.Examples include threaded discussion groups and bulletin boards, list-serves, archived video, and other anytime technologies.

Community College: The mission of the community college is to provide the necessary and appropriate combination of education (theory) and training (skills) to ensure that exiting students are prepared to be successful in the careers they are pursuing, or in the baccalaureate programs they are entering. This mission also applies to students currently employed in the field but engaged in professional growth and development and the pursuit of life-long learning.

Information Technology (IT): The field of IT is a large and ever-changing realm. Within this report we include the fields described in the ACM report *Guidelines For Associate-Degree Programs To Support Computing In A Networked Environment.* We also include the positions generally envisioned as "modifiers/extenders" (those who "modify or add on to an information technology artifact") and "supporters/tenders" (those who "deliver, install, operate, maintain, or repair an information technology artifact") as defined in the CRA report *The Supply of Information Technology Workers in the United States*. Excluded are the intended graduates of the Computer Science, Computer Engineering, and Software Engineering programs as defined in the ACM/IEEE Computing Curriculum 2001 draft report (http://www.cs.rit.edu/~spr/CLQABS/schneider.html).

Synchronous methods: Simultaneous (and typically scheduled) activities. Examples include chat rooms, videoconferencing, live video, and other real-time technologies.

Technician: A technician is someone who will enter the IT job market with either industry certification or a community college certificate, AAS, AS or AA degree. These technicians are qualified for a wide variety of jobs and each position may have a wide range of needed skills. We have attempted to define the base skills needed for these graduates, filling all of these positions.

APPENDIX B: Examples and Vignettes

Subnet Masking

Subnet masking enables the TCP/IP protocol stack to determine where to send packets of data. It is a method used to segment a network and give each segment a network ID, so that other networks can still communicate with it.

Segmenting the network greatly reduces traffic, because not all computers are trying to use the same bandwidth. However, it is sometimes necessary to send packets of information from one network segment to another, and in such a case a bridge or router must be used to combine the networks.

A router is a device that can distinguish the destination network ID of a packet sent on the network by using the destination IP address and subnet mask. It can then route that packet accordingly, without having to send it to all the other network segments. If a packet is not intended for a computer on a specific segment, the router will filter out that packet to reduce traffic on the network, and send it only to the segment that contains the destination computer's IP address.

IP Addressing

An IP Address is a binary address consisting of four eight bit numbers, or octets. These octets can be converted back and forth between decimal and binary notation.

You may be accustomed to an IP address such as 192.168.1.1, but the computer sees this address as 11000000.10101000.00000001.00000001. All IP addresses have a total of 32 bits, or digits that can assume either a 1 or 0 value. Even if all eight bits were ones, an octet can only add up to 255. This is why the octets in IP addresses never go above 255; 11111111 = 255 .

Each IP address has an affiliated subnet mask. All IP addresses on one network segment should have the same subnet mask. The subnet mask tells the computers and routers how many of the 32 bits describe the network identifier, and how many are left over to describe the individual computer (or host). Once the network identifier has been resolved, routers know to what network to send the packet.

A subnet mask basically "masks" the corresponding bits in a binary number. For example, a subnet mask of 255.255.240.0 in binary is

11111111.11111111.11110000.00000000

The masked bits will determine what subnet an IP address belongs to. The unmasked bits, or zeroes, are the unique address of the computer within that subnet. There are 20 masked bits in the number above. That means that for two IP addresses to be on the same subnet with a mask of 255.255.240.0, their first 20 bits must be the same.

If the subnet mask were the same as above, 255.255.240.0, then

11110110.10001001.10100111.10100110 (246.137.167.166) and
11110110.10001001.10101110.00010011 (246.137.174.19)

would be in the same subnet because the first 20 bits are the same.

Similarly, the numbers

00101001.11101111.00011111.10101010 (41.239.31.170) and
00101001.11101111.00010101.11111111 (41.239.21.255)

are in the same subnet for the same reason. However, they are in a different subnet than the previous two numbers.

To discover the Network ID, or the number that starts that particular subnet, set everything else to zeros. The Network ID cannot be used as an IP address, because it is all zeros.

11110110.10001001.10100000.00000000

Converting back to decimal, this becomes 246.137.160.0

The Boolean "AND" function can easily determine Network ID's. Simply take the subnet mask and "AND" it with the IP address, also known as the Host ID. For example, if the subnet mask is 255.255.248.0 and the Host ID is 199.199.69.2, then we have

11111111.11111111.11111000.00000000 (subnet mask)

AND

11000111.11000111.01000101.00000010 (Host ID)

which give

11000111.11000111.01000000.00000000 = 199.199.64.0, the Network ID.

Determining Number of Hosts and Subnets

Determining the number of subnets and the number of available host IDs per subnet is another computation that utilizes IP addressing and subnet masks and is routinely performed by IT technicians.

The number of subnets is determined by the formula 2^*n* – 2, where *n* is the number of masked bits, excluding those already predefined by the class type. For example, in the network ID of 192.168.1.0, with a subnet mask of 255.255.255.224, there are three masked bits counting from the Class C.

Provided Subnets = 2^3 – 2 = 6 subnets

The number of hosts/subnet is also defined by 2^*n* – 2, where *n* is the number of unmasked bits. In the preceding example, there are five unmasked bits.

Hosts/Subnet = 2^5 – 2 = 30 hosts/subnet

The total number of hosts = hosts/subnet * subnets. In this case, it is 30*6, or 180 total hosts.

Here is a typical problem an IT technician might face in this area. You need to divide a Class C network into 12 subnets. How many bits should you mask, and what will be the subnet mask?

Solution: Use the formula 2^*n* – 2. If *n* = 3, it provides for 6 subnets, which is not enough. However, if *n* = 4, it provides for 14 subnets, which is enough since not all subnets must be used. Because there are 4 masked bits, the subnet mask will be 11110000, or 255.255.255.240.

Trigonometry and Geometry

Provide the physical layout of an office, company, or building LAN, and determine the locations of network drops and cable lengths. Make efficient use of the cable to reduce costs (presuming approximately $1 per foot for cable). Avoid doors, lighting (RF interference), HVAC, etc.

This problem requires a student to explore the linear geometry of the space and to determine optimum (low use) cable lengths. Trigonometry can also be used to calculate such things as stresses on cable and cable runs when they are not properly supported. That is, to explore the vector analyses which can increase the load on a support member when weights are placed off center.

Geometry can be used to determine volume of network enclosures and cooling supply systems and to examine concepts of CFM, etc.

Relational Algebra

Within a table you need to query for results. These queries will need to create a relation between the entries in each table and create a cross product. The student should understand how to create the cross product and all potential errors that are possible.

Queuing Theory

Computer 1 (PC-1, with its own queue, Q-1), Computer 2 (PC-2, with its own queue, Q-2) and Computer 3 (PC-3, with its own queue, Q-3) are all requesting data from the Internet (sending data packets). These data packets are processed by a device called a router which has specialized software to deliver the data packets to the correct location. The router uses a queuing algorithm to process each of these packets, having its own queue to store all requests (Q-4). The problem occurs when PC-1, PC-2 and PC-3 create enough requests to fill both their own queues and the router's queue. The protocols and software on the router will halt all traffic from these queues until its own queue is stabilized.

Show the student how queues function and how the system will allocate resources to each of these systems and therefore cause the user to experience a slower response time.

Technical Mathematics:

Mechanical and Manufacturing Technology[1]

CRAFTY Curriculum Foundations Project
Los Angeles Pierce College, October 5–8, 2000
J. Sargeant Reynolds Community College, October 12–15, 2000

Al Schwabenbauer, John C. Peterson, and Kathy Yoshiwara, Report Editors
Bruce Yoshiwara and Gwen Turbeville, Workshop Organizers

Summary

For three days in October 2000, a team of community college faculty from mathematics departments and technical specialties met with industry representatives to define the mathematics requirements for entry-level technicians in manufacturing and mechanical technology. We also discussed our vision for the mathematics classroom of the future, including issues of theory versus application, instructional interconnections, methodologies, and delivery mechanisms.

We identified 13 mathematical skills areas in which students need a basic understanding. We also ranked 37 specific topics as high, medium, or low importance to students in this field. We recognize that degrees of specialization and differences among technicians in a manufacturing environment affect the ranking and depth of the topics.

We focused heavily on what employers expect from new technicians entering the workplace. Soft skills are very important: the ability to work in multi-disciplined teams and to communicate with other workers, engineers, managers, and customers. A technician must take real world data and information and use critical thinking skills to analyze a problem logically and formulate a solution. Troubleshooting equipment and processes is especially important in a manufacturing environment. A technician is also expected to use computer-based software for technical analysis as well as for communication and presentations.

We determined that a ratio of approximately 30% theory and 70% computation and application is the right mix in the classroom. We also agreed that students should be exposed to mathematics software and simulation, which are used in design and process planning and in statistical process control applications in business. All technicians should be able to use standard business software and the Internet for communication and presentation.

Our technical community college faculty felt that there should be major changes in the way college mathematics is taught today. The curriculum should be presented in a modular, just-in time format to suit the specific technical content area being taught. Mathematics problems must be realistic and relevant to

[1] This material is based on work supported by the National Science Foundation under DUE grant no. 0003065. Any opinions, findings, and conclusions or recommendations expressed in this paper are those of the authors and do not necessarily reflect the views of the National Science Foundation.

the technical field being studied. Team teaching, with mathematics faculty entering apprenticeships with technology faculty, would improve the teaching of mathematics concepts and applications required in the content area. Faculty should experiment with different classroom approaches to provide real world experiences for the students. For example, they might use the studio approach to integrate lecture and laboratory exploration and experimentation. In this environment they should provide for student team problem solving and for the development of leadership and communication skills. Both students and faculty should have the opportunity for internships and capstone projects with industry.

Narrative

Introduction and Background

Future Working Environment. In the coming decade, technicians in mechanical and manufacturing facilities will continue to perform a variety of jobs ranging from product and process design and planning through product manufacturing, testing, and delivery and field support. With increased globalization of markets and manufacturing processes, the focus will be on higher value-added tasks and on leadership in creating innovative products.

Laboratory technicians operate sophisticated measuring equipment for product development, testing and qualification, and for product conformity verification. In metal manufacturing facilities, NC (numerical control) machinists set up equipment, optimize process control feeds and speeds, perform statistical process control to maintain and ensure product quality, and perform preventative maintenance on their equipment. Other technicians perform sub-system and system level tests, verify that products meet all specifications, and provide trouble-shooting support for products in manufacturing as well as products already fielded with the customer.

Teamwork, communication and problem solving skills, within plants and work teams as well as across global networks of design and manufacturing facilities, will continue to be critical success factors for the individual technician and for US industry.

Technology Environment. The working environment for technicians in mechanical and manufacturing facilities in the next decade will call for increased computer skills, as both product and manufacturing processes incorporate more imbedded microprocessors for enhanced product functionality and real-time process control.

Technology should influence the teaching and learning of mathematics through the use of PC based modeling tools. Students should analyze real world environments with computer-based simulations and study multiple scenarios without having to perform tedious number crunching by hand.

Understanding and Content

Establishing a Baseline for Mathematical Content.

1. Teach students to identify the elements needed to set up an equation to solve a problem.
2. Decide at what level the use of calculators should be taught.
3. Equalize the preparation of incoming high school students. In some states high school students take their required mathematics in freshman and sophomore years, and then have a two-year gap before they take college entry mathematics tests. These students get discouraged in remedial mathematics classes, and a refresher class might be more effective. What do your incoming students need?
4. Demonstrate to high schools and two-year college students the importance of mathematics in general occupations and business today.
5. Train students to internalize mathematics skills. Application of real world problems would help with this internalization.

6. Recognize that the training for academic degrees (BS, MS, PhD) is completely different from the training for two-year technicians.
7. Provide bridge courses for job-oriented students in two-year programs to help them make the jump to a baccalaureate program.
8. Help advisors understand students' short range and long range plans.

What mathematical topics and content must students master during the first two years in order to complete their AAS program or to enter the job market?

We divide these topics in two groups. The first lists thirteen areas in which students should have a basic understanding. The second lists 37 specific topics and rates their importance.

Areas of Basic Understanding. All students at this level should have a basic understanding of:

1. Applied basic statistics for quality control and general business applications
2. Applied basic trigonometry
3. Applied solid geometry
4. Basic shop mathematics
5. Scientific and engineering notation
6. Conversions
7. Significant figures
8. Decimal to percentage conversion
9. Algebra: Quadratic equations, simultaneous equations
10. Theory behind setting up equations (modeling), problem definition
11. For technology students, mathematics, English, history, etc. should not be taught in isolation.
12. Construction and interpretation of basic graphs
13. Transference of data to information to analysis (critical thinking)

Topics Rated for Importance.
Topics are rated as low (L), medium (M), or high (H) importance to students in this field. (See the table on the next page.)

What mathematical problem solving skills must students master in the first two years?

Technicians need more training in critical thinking and analytical skills for problem solving. Application problems should therefore include critical thinking and analytical activities.

What is the desired balance between theoretical understanding and computational skill? How is this balance achieved?

The balance between mathematical theory (book learning) and computation or real world application should be 30% for theory and 70% for computation and application.

What mathematical topics are needed to advance up the career ladder and continue education to the bachelor's degree? What priorities exist among these topics?

1. Understanding of order of operations
2. Fundamental arithmetic
3. Converting fractions to decimals and decimals to fractions

1.	Mathematical models: development from verbal descriptions.			H
2.	Problem solving: application of multiple concepts.			H
3.	Integers: application in industry (differing concepts of zero)			H
4.	Decimal system			H
5.	Metric onversions – English to metric and metric to English			H
6.	Order of operations			H
7.	Ratios and proportions			H
8.	Percentages			H
9.	Approximation			H
10.	Linear measurement			H
11.	Powers and roots		M	
12.	Exponents		M	
13.	Logarithms (as related to areas in electronics, friction, etc.)			H
14.	Angles			H
15.	Radians (as conversions)			H
16.	Decimal conversion to degrees/minutes/seconds and back	L		
17.	Geometric relationships (not geometry proofs)			H
18.	Bisection of a line	L		
19.	Determining the center of a circle	L		
20.	Inscribed and circumscribed circles	L		
21.	Triangles, both right and oblique			H
22.	Pythagorean theorem			H
23.	Basic trigonometric functions: sine, cosine, tangent			H
24.	Variables			H
25.	Solution techniques for linear equations			H
26.	Simultaneous equations; application of multiple methods			H
27.	2 D or general graphing skills			H
28.	Quadrants: I, II, III, IV		M	
29.	Polar coordinates: introduction			H
30	3 dimensional coordinate systems: x, y, z		M	
31.	5 axes & 6 axes: x, y, z, a, b, c	L		
32.	Rates of change, conceptually understood (not requiring calculus)			H
33.	Vectors: components, addition, subtraction		M	
34.	Basic statistics as applied to industrial quality assurance: mean, median, mode, standard. deviation, X-Bar & R, Range, skewness, average, trends			H
35.	Pareto plots, histogram, scatter plots			H
36.	Detecting trends in data sets			H
37.	Interpolation and extrapolation			H

4. Basic algebraic equations
5. Application of skills—especially important
6. Approximation and estimation: having a feel for the right order of magnitude, the right units of measure, and the appropriate precision for an answer
7. Cartesian coordinates

Technology

How does technology affect what mathematics should be learned in the first two years?

After introducing a topic or concept (such as linear equations) move immediately to the technology to speed computation in practical problems. Use technology to overcome tedious number crunching and allow consideration of many problems, introducing students to the real world environment of solving multiple problems.

Use computer-based modeling or simulation software. In the not too distant future every student may have access to a laptop. Laptops could actually be integrated into the classroom, especially in technology courses.Link mathematics software to application software in order to reinforce mathematical concepts.

What mathematical technology skills should students master in the first two years?

Students need to be computer and calculator literate. This means our students should have experience with graphing calculators, word processors, spreadsheets, data base management software, and computer presentation applications such as *PowerPoint*. They should also have experience with mathematical and statistical software such as *Maple, Mathematica, Mathcad,* or *Statistica.*

Instructional Techniques

What are the effects of different instructional methods in mathematics on students in your discipline?

The "studio approach," employed in some mathematics courses, may better integrate lecture, lab, exploration, experimentation, etc., for our students. Cross-departmental team teaching has the potential to present mathematics in an applied context that can be more effective with students in technical fields. The use of collaborative problem solving teams reinforces important workplace skills of cooperation and team work.

What instructional methods best develop the mathematical comprehension needed for your discipline?

For our students the best instructional methods are those that use real life hands-on applications, repetition, case studies, and computer simulations. Avoid teaching mathematics to simply teach mathematics. Move heavily to the application of the concepts within the mathematics classroom. Students gain much when instruction is coupled with internships, cooperative work assignments, or work shadowing arrangements.

Instructional Interconnections

What changes are needed in the mathematics curriculum in order to satisfy the needs of AAS students and technicians?

The traditional mathematics curriculum includes College Algebra, College Trigonometry, and Intermediate Algebra, all of which are set up to support the mathematical needs of the college as a whole. Technical mathematics courses are tailored to be program specific, and there are significant variations in such courses among the community colleges that offer them.

Devise curricula that provide mathematics instruction in a just-in-time format. When topics are needed in technology, physics, or science, the mathematics topics would be coordinated for delivery. We understand that this will not be easy. Faculty members will need to be extremely cooperative with each other. Mathematics instructors may need to apprentice in technology areas to learn the applications and understand which mathematics topics are important and when these topics are needed.

What instructional methods might mathematics instructors use to develop or reinforce non-mathematical skills or understandings in your discipline or company?

1. Team teaching
2. Emphasizing critical thinking
3. Demonstrating a logical approach to solution of a problem
4. Requiring report writing and presentation (communication)
5. Just-in-time teaching: teaching each mathematics concept and application as it is required in the content areas
6. Working together with technology colleagues to identify expectations, trying different approaches, adjusting the delivery to focus on desired outcomes
7. Including problems with a range of answers instead of a single answer
8. Using student teams on problem solving applications. Use of team problem solving techniques should begin as early as possible, because the skill develops over time.
9. Having students present *how* they solved a problem.

How can dialogue on educational issues between your discipline and mathematics best be maintained?

1. Increase and improve communication between faculty members in mathematics and the technical fields.
2. Integrate the mathematics department into the content area programs.
3. Allow better assessment of the outcome of classes in mathematics, English, and government by technology faculty. These departments are really in a support role, as opposed to classical or traditional mathematics departments.
4. Make sure that mathematics and other support curricula are not delivered in a vacuum, divorced from the technical content areas.
5. Recognize that the mathematical support of technical areas should differ from the mathematical preparation of transfer (academic) students. Change can be incremental: find one sympathetic mathematics faculty member, then a second, then a third, etc.
6. Establish an ongoing dialogue with industry representatives. Encourage industry participation on advisory committees.
7. Initiate faculty shadowing in industry, coops, and similar programs. These activities should be ongoing, not just every 3, 5, or 10 years.
8. Invite mathematics and English faculty to serve on technology advisory committees as active participants.
9. If possible, solicit student evaluation of their educational experience after their entrance into the workplace. (Some students or companies may be reluctant to do this.)
10. Establish dialogue with the high schools.

WORKSHOP PARTICIPANTS

Ashok Agrawal, Department Chair and Professor, Plastics Technology, St. Louis Community College, St. Louis, Missouri

Debra Cohen, Coordinator of Assessment, Research & Planning, John Tyler Community College, Chester, Virginia

Larry Earnhart, Technical Leader, Performance, Polymers and Chemicals, Honeywell International, Inc., El Segundo, California

Mike Farley, Marine Designer, National Steel and Shipbuilding (NASSCO), San Diego, California

Reece Gibson, CNC Instructor, Metropolitan Community Colleges, Kansas City, Missouri

Nick Johnson, Technical Instructor, Bosch Automotive, Anderson, South Carolina

Tracy Koss, Engineering Manager, Marathon/Ashland Corporation, Findlay, Ohio

Jack Mclellan, Quality Assurance Coordinator, Mott Community College, Flint, Michigan

Greg Meserve, New Hampshire Technical College, Berlin, New Hampshire

Bob Monter, Technology Specialist, Wright Technology Network, Dayton, Ohio

John Reed, Training and Development Specialist, Exxon/Mobil, Baton Rouge, Louisiana

Frank Rubino, Middlesex Community College, Edison, New Jersey

Al Schwabenbauer, Vice President, Sikorsky Aircraft, United Technologies Corporation

Jim Shimel, Machine Tool Instructor, Metropolitan Community Colleges, Kansas City, Missouri.

Emin Turker, Dean of Engineering Technologies, Lakeland Community College, Mentor, Ohio (Currently Dean of Engineering Technologies, SUNY Canton, Canton, New York)

ACKNOWLEDGEMENTS

The members of the panels would like to thank Bruce Yoshiwara of Los Angeles Pierce Community College and Gwen Turbeville, of J. Sargeant Reynolds Community College for coordinating the workshops at their respective campuse. We would also like to thank the Mathematical Association of America for initiating this series of workshops and to The American Mathematical Association of Two-Year Colleges for providing this opportunity to consider and make recommendations on a topic of importance to the information technology community; especially project directors Mary Ann Hovis, Rob L. Kimball, and John C. Peterson. Finally, we would like to thank the National Science Foundation for providing the funds to support these meetings.

APPENDIX A: Examples and Vignettes

1. Machine shops today are moving to multi-axis, NC programmable equipment and machining cells, including very accurate CMM (Coordinate Measurement Machines).

   The operator/technician is expected to be able to setup this equipment, optimize the routing feeds/speeds for different materials, perform total productive maintenance (TPM) on the equipment and participate in Total Quality Management (TQM) such as Six Sigma Quality Management Tools, practiced in most U.S. Manufacturing companies today.

   Skills for set-up, NC optimization:

   Geometry, trigonometric functions, solving equations, measurement and conversion skills.

   Skills for Total Quality Management

   Statistics: mean, standard deviation, UCL, LCL,

   Histograms, scatter diagrams, average, range, etc.

   Other Skills expected or required in a modern manufacturing company:

   Communication skills: written, oral, and computer (PowerPoint)

   Leadership skills

   Ability to work on teams, to be multi-disciplinary

   Ability to think logically, and to troubleshoot equipment and processes

2. Find the tension forces in cables *AC* & *BC* as shown in the following diagram:

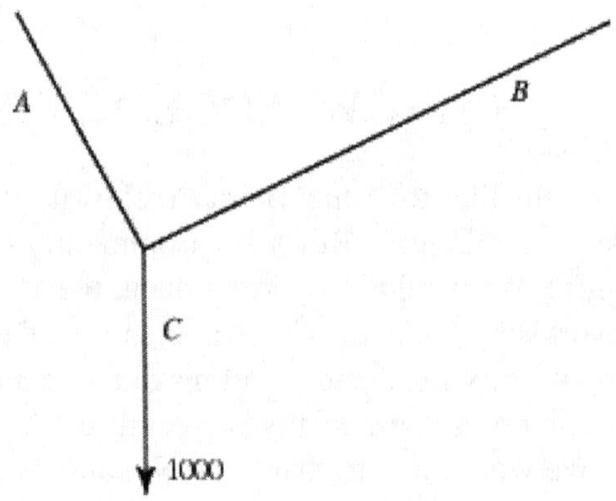

   a. Draw the free body diagram.
   b. Place the vectors tip to tail to form a force triangle.
   c. Use the Law of Sines and Law of Cosines to find the force in the cables.
   d. Find the components of the unknown forces in Cables *AC* and *BC*.
   e. Using the components of the forces and the applied force, write the equations for

$$FX = 0 \text{ and } FY = 0$$

   f. Solve the system of equations using a calculator and determine the force in each cable.

3. Teaching of Frequency Distribution: example with Alpha Bits. (Boxes made in different factories do have different frequency distributions. Marshmallow Alpha Bits have a different frequency distribution than "standard" Alpha Bits.)

a. Population: the contents of the box
b. Sample: 100 "letters" in the sample
c. Tally sheet: shows the frequency of each letter
d. Lot traceability: production lot on the box
e. Graph: the frequency distribution

4. 2D Graphing—using a formula to create an involute profile needed to produce a two-dimensional CAD drawing of spar gears.

5. Combination of strength of materials, manufacturing, and mathematics: design a container to hold and protect two eggs. The container should be able to protect the eggs when dropped off of a specified building. Build the container and drop it off the building: do the eggs survive? (No, you can't hard boil the eggs!) Other options and concepts can be added, such as:
   Assemble the container using at least two assembly methods.
   Assemble the container using at least two different materials.
   Use trigonometry to calculate the height of the "drop" building.
   Calculate the terminal velocity of the egg container.

6. Piping River Crossing. Schedule 80 pipe, 10 inches in diameter, is to be laid across this mile wide, fresh water river. The current is negligible and the river depth reaches 40 feet. The question: determine if the pipe will float when it is empty. Further questions:
   What is Schedule 80 pipe? (Internet research)
   Change to a salt-water river (sea water). The water density changes.

7. Building a fire sprinkler system: pipe pressures. How much pump pressure (head) PSI is required at ground level to operate fire sprinklers on the 8th floor of an office building? The following is the given information. Each story of the building is 12 feet high. Each sprinkler head is 9 feet above the floor, and the required pressure at the sprinkler head is 10 PSI. The length of the pipe is 300 feet and the friction loss in the pipe is 1.2 feet of head per 100 feet of pipe.

8. Swimming pool filling—volume, etc. Given the dimensions of a swimming pool, a hose delivering water at a specified rate (gallons per minute), and the cost of water per gallon, determine how long it will take to fill the pool and how much the water will cost. This will involve a conversion of cubic feet into gallons.

9. Ratios. Problems can involve 2-cycle engine oil ratios, cutting fluid ratios in machining situations, weed killer or yard fertilizer ratios. Problems centered on snow blowers and chain saws can also be devised. Or octane boosters for cars.

10. Automotive problems. Can involve miles per gallon, Kg/Gal, maintenance and operation costs.

11. Problems involving "grocery store mathematics" or "Home Depot mathematics" can be useful. These can involve converting fractions into decimals: "You and seven friends at the Brunswick Wild Oats Bakery have a $10 bill—how much can each of you spend?"

12. Proportions: using recipes such as in cooking or mixing cement.